ISBN 978-0-265-17501-9
PIBN 10169088

FB &c Ltd, Dalton House, 60 Windsor Avenue, London, SW19 2RR.
Company number 08720141. Registered in England and Wales.

For support please visit www.forgottenbooks.com

A

HISTORY

OF THE

FISHES

OF THE

BRITISH ISLANDS.

BY JONATHAN COUCH, F.L.S.

VOL. III.

CONTAINING FIFTY-NINE COLOURED PLATES,

FROM DRAWINGS BY THE AUTHOR.

The works of the LORD are great, sought out of all them that have pleasure therein.—PSALM cxi, v. 2.

LONDON:
GROOMBRIDGE AND SONS, 5, PATERNOSTER ROW.
M DCCC LXIX.

CONTENTS.

ATHERINA.

The body moderately lengthened, and along each side a silvery band from the gill-covers to the tail. Mouth a little protractile. Dorsal fins widely separated, both with rays. Ventral fins behind the pectorals, (abdominal.)

Linnæus confounded all the species together under the name of *A. hepsetus*, which, as now defined, is a species not known in Britain.

ATHERINE.

SMELT. SAND SMELT. In the west of Cornwall, GUID, which signifies white.

Atherina,	JONSTON; p. 79, table 17, f. 15.
"	WILLOUGHBY; p. 209, table N. 12, f. 2.
Atherina presbyter,	CUVIER.
" *hepsetus,*	DONOVAN; pl. 87.
" "	FLEMING; Br. Animals. p. 217.
" *presbyter,*	JENYNS; Manual, p. 377.
"	YARRELL; Br. Fishes, vol. i, p. 229
"	GUNTHER; Cat. Br. M., vol. iii, p. 392.

This little fish is commonly found in small scattered companics in harbours and bays where a sandy bottom is mingled with rough ground, and especially where there are streams ot fresh water flowing into the salt; but we believe there is no instance of its entering fresh water for any continued length of time, and it appears never to go to a considerable distance from the shore.

It inhabits the temperate regions of Europe, its most southern limits being the Island of Madeira. In the Mediterranean it is common, and is perhaps the species mentioned in a cursory manner by Oppian. It is also more abundant on the west and south of England than in the north or in Scotland, where, however, in the Firth of Forth, it was discovered by Dr. Parnell; so that we conclude its not being met with on the east coast of England, northward of Dover, is rather caused by the nature of the ground than the coldness of the climate. Yet there is reason to believe that this fish is in a high degree susceptible of the influences of cold; for besides that it goes into deeper water or more sheltered places in winter, I have been informed of instances where, in the shallow waters of a harbour, numbers have been surprised by sudden frost, so as to be left dead on the shore. Atherines sometimes assemble about the ends of piers, where they take a bait readily; but this would be tedious work for a fisherman who labours for profit, and they are usually sought after with a net formed of fine twine, by which large quantities are sometimes caught. Three small boat loads have been taken at a single haul.

The pretty look of this fish, and its resemblance at first sight to the true Smelt, *Osmerus eperlanus*, although of a different family, has secured for it a ready sale at our fashionable watering-places; but among writers on natural history there is much diversity of opinion as regards its excellency for the table. By some it has been highly prized, especially when large with roe, and fried without removing the entrails, and we have Montagu's authority for its excellency in this condition; but it has been so decidedly rejected by others as to have afforded cause for supposing that even its name of Atherine has been given it on account of its worthlessness. It seems certain that when out of season the multitude of its small-pointed bones are a material drawback from the pleasure of eating it.

I have found the roe enlarged into the full size from the middle of March to the end of August; and the spawn is contained in a single lobe, which is enclosed in a black covering of peritoneum, and has its origin near the backbone, from which it passes forward across the intestine, a little before the vent.

The air-bladder is not situated far forward, but is rather large, and passes beyond the vent towards the tail. In collecting them for observation I have found many more females than males.

Mr. Thompson gives the greatest length of an Irish example as seven inches and three fourths, and a Cornish specimen has measured but a little short of this, but a usual size is five or six inches. The form of the body moderately lengthened and slender; round over the back, so as to be only in a moderate degree compressed on the sides; the body covered with large scales; lateral line scarcely to be distinguished, but passing along the middle of the characteristic silvery stripe, which, in a fish of the larger dimensions given above, was one fifth of an inch wide, the body being an inch in depth, and this stripe runs straight from the origin of the pectoral fin to the tail. There is an appearance of transparency in the body, but this is an optical deception, and no inward part can be discerned from without. The head is flat above, but compressed at the sides; eyes large and lateral, the snout short before them; upper jaw arched; under jaw, seen from below, shaped like a horse shoe; angle of the mouth depressed; teeth small and numerous. Vent about the middle of the body, as measured from the snout to the fork of the tail. Dorsal fins two, separate, the first beginning a little behind the end of the pectoral, having eight rays, the first of which is at the centre of gravity of the fish. Second dorsal and anal opposite each other, the first with thirteen and the second with seventeen rays; anterior portion of these fins elevated, and the posterior margin narrow. Pectoral fin high on the side, superior rays longest, fourteen in number. Ventral fins with six rays. Tail divided, with eighteen rays. Colour a greenish transparent grey above, sometimes tinged with pink near the silvery stripe; transverse bands or lines across the scales on the back. I have seen the second dorsal, anal, and caudal fins dotted over with small round spots.

BOIER'S ATHERINE.

Atherina Boieri, RISSO. CUVIER.
" " Report of Penzance Society of Natural History, for the year 1849.
GUNTHER; Cat. Br. M., vol. iii, p. 394.

IN the autumn of the year 1846, in the midst of turbulent weather, there was discovered in the harbour of Polperro, on the south-east coast of Cornwall, a large number of small fishes, which manifested actions that attracted attention as being unlike those of species commonly known. Their first appearance was traced to the 21st. of October, and as the roughness of the sea became calmed down it was ascertained that they were a species of Atherine; but to obtain an example became a matter of no little difficulty.

The larger sort of Atherine, already described, (*A. presbyter*, or Smelt,) usually swims at a considerable depth in the water, but in the present instance they were all near the surface, not more than three or four within a foot or two of each other, but the whole scattered loosely over the water, to the number of several thousands. Their heads were in one direction, as if passing inward, and they were constantly rising dimples on the surface, like scattered drops of rain, by apparently examining or seizing some floating object; but, however earnestly engaged, their vigilance was never remitted, and it became scarcely possible to approach them, as in an instant they were off in another direction at the sight of a moving object. As it was found that their mouths were small—as indeed might be expected in fishes which did not exceed three inches in length—it was difficult to find a hook which they could take; and when a bait was offered, it was seen that they would not notice it until it was made to assume some of the actions of moving life; and it was by doing this that a few examples were secured. Seating myself on a projecting rock, near which they were passing, it was found that not one would come near it as before, and it was only by close concealment and a fortunate dip of a net that a few others were obtained. They continued thus for a week, at the end of which they had disappeared, and, although carefully looked for, they have not since been seen.

The examples then obtained were conveyed to London, and, by examination of Cuvier's "History of Fishes," in the library of Mr. Yarrell, no doubt existed that they answered to Boier's Atherine of that work, and consequently of Risso, who first gave it that distinctive name. The specimens, preserved in spirit, were presented to Mr. Yarrell, but they are not found in the collection of that gentleman now in the British Museum.

As the whole of those which came under observation appeared of one size, and those which were caught measured in length about three inches, we may suppose this to be their usual size; but it differs from the more common Atherine in the proportionally larger eye, more projecting lower jaw, with the dorsal fins nearer together. The colour in each was much alike. According to Dr. Gunther the roe is in a single lobe, as we have noticed in the last species

MUGIL.

The body stout, compressed at the sides, covered with large firm scales. Head arched across; the mouth wide, with a narrow gape; the lips fleshy; teeth exceedingly fine; middle of the lower jaw bent up and received into a recess of the upper. Dorsal fins two, separate, the first with a few spinous rays. Ventral fins abdominal.

As the close inquiries of Dr. Gunther into the several species of this family of Mullets appear to shew that there is a larger number of sorts on our coasts than naturalists have hitherto supposed, it will require on the part of an observer no small amount of discrimination to come to a conclusion concerning the examples which may come in his way; and any inquiry on the subject, to be satisfactory, will demand that any doubtful example shall be fresh from the water, or, if that be impossible, the specimens must have been preserved in a better manner than is usual with fishes. Our history and description of these Mullets will comprise, in the first place, those kinds with which we are acquainted, and after this we shall have recourse to that information with which the kindness of Dr. Gunther has supplied us, in assigning those distinguishing marks by which these little-known species may be recognised. The habits of these obscurer kinds, so far as they may differ from the others, are still unknown.

These fishes are frequently called by the name of Grey Mullets, to distinguish them from the Red or Surmullets, from which they differ in almost every characteristic of form, colour, and habits.

GREY MULLET.

GREAT MULLET. MULLET.

Mugil,	Jonston; pl. 23.
"	Willoughby; p. 274, table R. 3.
Mugil cephalus,	Donovan; pl. 15.
" *capito,*	Cuvier.
" *cephalus,*	Fleming; Br. Animals, p. 217.
capito,	Jenyns; Manual, p. 374.
"	Yarrell; Br. Fishes, vol. i, p. 234.
" "	Gunther; Cat. Br. M., vol. iii, p. 439.

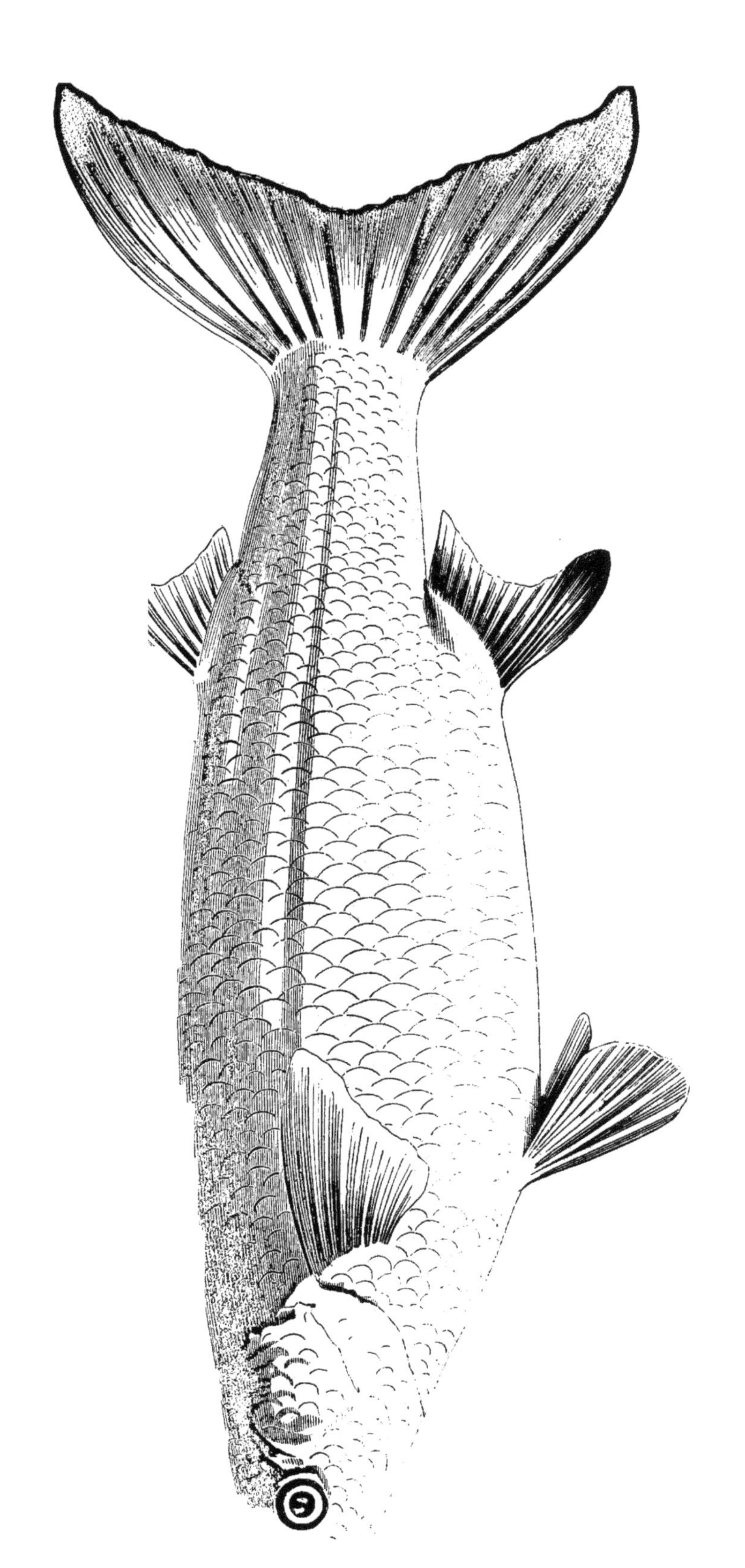

The Grey Mullet is familiarly known round the coasts of the United Kingdom, although perhaps it exists in greater numbers in the south and west than in the north. It appears indeed to be more widely distributed than others of its genus, for while it is met with in the north of Scotland, Ireland, and Scandinavia, it is more abundant than even with us through the Mediterranean, even to the mouth of the Nile; and southward it is known at the Cape of Good Hope.

Its choice of residence is in the neighbourhood of the shore, where it is most frequently seen in harbours, especially where the larger rivers empty themselves into the ocean; and this fish is even known to leave the salt water altogether for an occasional change, although by choice it soon returns to the waters of the sea. From a study of the younger individuals there is reason to conclude that this interchange between the waters of the sea and river is of importance to their health and growth, but as regards the fish of full growth the change may be especially beneficial by affording some supply of favourable food, as it has grown into a proverb that an Arundel Mullet is of particularly delicious quality. When living in a muddy bottom or foul water, as is the case in some parts of the Mediterranean, as food it is said to lose much of its admired flavour. We are told that a gentleman who kept Mullets confined in a pond of fresh water, where no doubt they were sufficiently supplied with food, was persuaded that by this treatment they were much improved in quality as well as growth; but how long this exclusive confinement was persevered in does not appear.

From Columella we learn that they were among the fishes preserved in ponds by the ancient Romans, but these we suppose to have contained the water of the sea. We are left by this author to suppose that these fishes did not suffer by the confinement, and from the number he mentions it seems probable that they even increased in their captivity. This Roman author, whose observations probably were drawn from Mullets thus confined in ponds, applies to them the epithet of sluggish, *(iners mugil,)* but this can only be appropriate to the deliberate manner in which the structure of their mouth compels them to proceed in taking their food; in which it must be allowed they are much more slow than the Bass or Lupus, which he mentions immediately after them, and against which they were believed to feel a perpetual antipathy. The charge of imbecility brought against this fish by Pliny, as shewn by the fact that it hides its head for concealment when alarmed, and then acts as if persuaded that its whole body was concealed, is, as Cuvier has remarked the opposite to what we know of the character of these fishes; of which the vigilance, when exposed to observation, is very great, although this is aecompanied with little appearance that might lead us to suspect its existence.

In the desire for food, which is a predominant appetite in the generality of fishes, the Mullet appears to shew itself fastidious; but this appearance arises from the fact that from natural causes its range of choice is limited, and of no other kind of fish can it be so safely affirmed that it rarely selects anything for subsistence that is endued with life. Such also was the opinion of Oppian:—

> "Mullets, unlike the rest, are just and mild,
> No fish they harm, by them no seas are spoil'd;
> Not on their own nor different kinds they prey,
> But equal laws of common right obey.
> Undreaded they with guiltless pleasure feed
> On fattening slime, or bite the sea-grown weed.
> Each licks his mate———"

It must be confessed, indeed, that this last particular is not literally true, for the Mullet will devour a worm when presented to it, and it is even fished for successfully with a fly; but from Mr. Thompson's account of the habits of the Mullet, as he described them in his "Natural History of Ireland," vol. iv, some doubts may be felt whether he speaks of the same species. He says:—"The contents of the stomachs I have examined at various seasons, presented (from the minute size of the objects,) many hundred-fold greater destruction of animal life than I have ever witnessed on a similar inspection of the food of any bird or fish. From a single stomach I have obtained what would fill a large-sized breakfast-cup, of the following species of bivalve and univalve mollusca, which had been taken alive,—*Mytilus edulis*, *Modiola papuana*, (of these very small individuals,) *Kellia rubra, Skenea depressa, Littorina retusa, Rissoa labiosa,* and *R. parva, serpula,* and *miliolæ.* Of these mollusca, specimens of *Rissoa labiosa* three lines in length were the largest, and the *Kellia rubra,* from the smallest size to its maximum of little more than a line in diameter, the most abundant. In the profusion of specimens it affords the stomach of one of these Mullets is quite a storehouse to a conchologist. In addition to these there were various species of minute crustacea. The only inanimate matter that appeared were fragments of *Zostera marina* and conferva, which were probably taken into the stomach on account of the adhering mollusca. To this nutritious food may perhaps be attributed the great size this fish attains in Belfast Bay." Mr. Thompson adds, that in the "Animal Kingdom" of Cuvier, Pennant's figure of the Grey Mullet in his "British Zoology," is referred to as *M. capito,* but in the "History of Fishes," by Cuvier and Valenciennes, it is believed to represent *M. chelo.* In this last work Donovan's figure of the Mullet, pl. 15, is considered a very good representation of *M. chelo,* although Yarrell and Jenyns refer to both figures as *M. capito.*

I will here add that in the references I have made, as in the history given, my opinion respecting the species is the same as that of Yarrell and Jenyns, and that our history applies only to the identical species we have described and represented. It has been remarked by different observers that this fish is sometimes, and perhaps often, seen to grope in the soft flooring of the bottom, with the help of its very sensitive lips and curiously-formed mouth, by which every particle is closely examined, and to swallow a mixture of decaying vegetable and animal substances, with sand, of the latter of which alone I have obtained from a single stomach so much as would fill a tablespoon; but no one will suppose that the sand so swallowed was a principal object of search. The very minute mollusks mentioned by Mr. Thompson would at least be as acceptable as the sand, and without doubt much more so; but that this Mullet can live and thrive where such food is beyond its reach is out of the question.

No shell or substance beyond the size given by Mr. Thompson can pass into the stomach of the Mullet, for, after the close sensitive examination it has undergone at the entrance of the mouth, it has to be strained through a sifting apparatus in the throat, by which all that is stout or rough becomes rejected, and then blown out of the mouth; after which what forms the food is received into the firm and muscular stomach, that for substance resembles the gizzard of a fowl. By the action of this organ what is capable of affording nourishment becomes digested, and the remainder is passed on through a thinner portion of the stomach to be finally expelled at the vent.

Oppian has taken notice of a delicate trait in the character of this fish, in an action which however is sometimes noticed of other species when not very eager for food, and which action is also mentioned by Ovid:—

"The scenting Mullet creeps with slow advance,
And views the bait with coy-retorted glance.
First with his tail he feels the bait, and tries
If vital warmth the beating pulse supplies,
For Mullets always spare the living prize;
Then slightly nibbles, but perceives too late
The doubted fraud, and feels the pungent fate."

The form of the mouth and narrowness of the gullet form a hindrance which prevents this fish from swallowing a hook of even small size; and generally the close examination which is made of the texture of what is to be admitted might appear a sufficient guard against the reception of anything that might endanger its safety; yet the Mullet is not unfrequently caught with a line, and the misfortune itself is the result of those very actions which seem best fitted to ensure its safety. The close pressure of the lips on the bait will cause the point of the hook to pierce the flesh, and in this way the fish falls a victim of mischance, when, however, no slight skill and patience are required to bring the prize safely to land. The baits employed are a small soft worm, some fatty substance, or cabbage boiled in animal broth; and Oppian mentions, as an ordinary bait in his day, a mixture of curds of milk with flour and an infusion of mint, fastened on an ordinary hook.

But the Mullet is more frequently sought for with the net, and it is in its encounter with this that the large amount of its watchful intelligence and activity are displayed. As we are informed that this fish is an inhabitant of the Mediterranean through its whole length, and, according to Mr. Frazer, "Travels in the Persian Provinces," even in the Caspian Sea, we feel justified in believing that it is the identical species, in common perhaps with the *M. cephalus*, a kindred fish not yet discovered in Britain, of which Oppian has described the proceedings when beset with dangers; and it is with pleasure we are able to adduce instances of the accuracy of the representations that are given of these by the poet, who has referred to this fish more frequently perhaps than to any other, and always with much discrimination of character. The net, of course, is shot near the shore, where its leaded bottom may reach the ground; and the intention of the fisherman is, either to let it remain moored until the tide has ebbed away, or at his convenience to draw it to land. But the watchful eye of the fish will often discover the snare even before the barrier of restraint is altogether closed. Instead, however, of rushing to the only but yet distant opening, which it might not be able to reach in time, it boldly essays to spring, or rather tumble over the head-line, and when one succeeds all the rest will be sure to follow.

> "The Mullet, when encircling seans enclose,
> The fatal threads and treacherous bosom knows.
> Instant he rallies all his vigorous powers,
> And faithful aid of every nerve implores;
> O'er battlements of cork updarting flies,
> And finds from air th' escape that sea denies.
> But should the first attempt his hopes deceive,
> And fatal space the imprison'd fall receive,
> Exhausted strength no second leap supplies;
> Self-doomed to death the prostrate victim lies,
> Resigned with partial expectation waits
> Till thinner element completes his fates."

This last particular is also emphatically true as regards another method in which escape is attempted; and I have seen where all its companions have gone over the corks, and the head-rope of the net has been raised high above the water, a solitary prisoner has examined all the meshes through the length and breadth of the net, and finding no opening through which it could readily pass, it has retired to the greatest distance the space admitted, and thence with a rapid dash has sought to thrust itself through. This last desperate effort has been in vain, and from that moment the fish resigns itself to its fate.

So strong also is this impulse of watchfulness against restraint, that to avoid it the Mullet will encounter danger, even when the space enclosed is of considerable extent. In the port of Looe, in Cornwall, there is a salt-water mill-pool of thirteen acres that is enclosed on the side of the river by an embankment, and into which the tide flows through flood-gates that afford a ready passage for fish to the space within. When the tide begins to ebb the gates close of themselves, but even before this has happened the Mullets which have entered have been known to pass along the enclosed circuit within the bank, as if seeking the means of deliverance, and, finding no outlet, they have thrown themselves on the bank at the side to their own destruction. Even Mullets of exceedingly small size have been seen to throw themselves, head or tail foremost, over the head-line of a net, where it would have seemed much easier for them to have passed through a mesh; and so strong is this propensity to pass over an obstacle rather than through it, that examples of less than an inch in length have repeatedly thrown themselves over the side of a cup where the water was an inch below the brim. Fishermen, however, are acquainted with a simple method, which, by deceiving the fish, is sufficient to prevent their taking a successful leap over the net. A thin layer of straw is scattered over the surface to the breadth of a few feet within the head-line; and mistaking this for the obstacle itself, the fish exhausts its efforts on the wrong object, and remains a prisoner still.

Risso describes another mode of taking this fish, by attracting it with a light, and then darting at it a spear or trident,—perhaps the crossed trident, or such as by sailors in England is termed the grains; but it scarcely appears successful with us, although ingeniously contrived for the purpose.

But a more remarkable and singular method of taking Mullets is mentioned by ancient writers, although with some variation as regards the particular species of Mullet. Pliny (B. 9, C. 25) refers to it as simply the Mugil, the salacious properties of which render them so unguarded, that in Phœnicia, and also in the province of Gallia Narbonensis, at the time of coupling, which is about midsummer, and near the influence of fresh water, an individual of either sex, which was taken out of the preserved pond, was fastened to a long line that was passed through the mouth and gills, and then the fish was left to wander to the end of the line in the sea; after which it was drawn back again, when it was found to be followed to the water's edge by some one or more of the opposite sex. Ælian relates this more at large, and says the decoy fish must be selected as the most excellent and beautiful of its kind; but in the title to his account of it he refers this habit to a species he terms *Oxyrhynchus kephalus*, or Sharp-nosed Mullet, which he appears to distinguish from the simple Kephalus, as also from the Kestreus, which is another kind of Grey Mullet. Oppian also makes a distinction between the Kephalus and Kestreus, but refers to the same practice of attracting the Kephalus near the shore, where a casting-net was thrown to secure the prize. And strange as this story is, it is borne witness to by Gesner, who is quoted by Willoughby as having seen it practised at Tarentum. A male fish was observed to follow a female that had been sent out as a decoy, and, although severely wounded with a spear, it would not be made to quit its lure, until at last it fell a victim to its love.

All writers agree in ascribing to this fish great quickness of hearing, and it has even been supposed that it is capable of the perception of particular sounds. The Cornish historian Carew had formed a pond on a branch of the Tamar, in which Mullets were fed at regular periods, and they were drawn together to the appointed spot at the sound made by the chopping of their food. We are not to conclude it certain that the sacred fishes mentioned by Martial, as being preserved in the Roman emperor's pond at Baiæ were Mullets, although it is probable they were so; and it may have been with some exaggeration that he says they each one knew their name, but, where the sound was simple the general observation of the fact is not without probability.

From all accounts, ancient and modern, it is certain that this fish has ever been in esteem for the table, although in some places more than in others; and Ausonius says that to be eaten in perfection it should be cooked within six hours after it is caught. But there is a favourite preparation made from it in Italy that is scarcely known in England. It is called botargo, and is formed of the roe: which is carefully removed from the fish, and sprinkled with salt for four or five hours, after which it is pressed between two boards and dried in the sun by day for thirty or forty days, or by some it is dried in smoke. It is supposed to sharpen the appetite, excite thirst, and heighten the relish of wine.

This fish grows to the length of eighteen or twenty inches, and will sometimes weigh from twelve to fifteen pounds. The body thick and solid, but compressed at the sides, the head wide and flat on the top, compressed on the cheek. In one that measured eighteen inches in length the greatest depth was four inches. Eye moderate, lateral, round, in a line with the angle of the mouth. The gape narrow; jaws equal, the lower bent up at the middle to form a keel, which is received into a cavity in the upper jaw; both jaws are capable of some degree of extension; the teeth so fine that they are not always to be discovered, hair-like, closely set, with their points set in a crenated line. Lips membranous or fleshy, with raised fleshy lines in two rows, except at the symphysis of the upper lip; a slight roughness on the tongue and a small portion of the vomer. The mystache or maxillary bone separate from the fleshy lip, turned back near the eye. The body covered with firm scales, which extend over the cheeks. First dorsal fin begins at about midway between the upper lip and root of the tail, with four firm rays; second dorsal removed from the first by more than the length of the former, with nine rays, the first simple and slight. Anal fin opposite the second dorsal, and rather longer, with eleven rays, of which the third is the longest. Pectoral fin broad and high. with sixteen rays. Six rays in the ventral fin, the first simple. Caudal incurved, with fourteen rays, besides two or three false rays.

The colour on the back is a dark bottle green, which, when out of the water soon fades into grey; lighter on the sides and belly, with broad lines of a deeper colour running towards the tail, varying, but about seven in number, cheeks and border of the pectoral fin tinged with yellow; iris of the eye dark brown. Large well-marked facial nerves pass forward near the angle of the mouth, to be distributed and afford special sensation to the upper lip and its raised lines.

LESSER GREY MULLET.

Mugil chelo, JENYNS; Manual, p. 375.
" " YARRELL; Br. Fishes, vol. i, p. 241, but Dr. Gunther represents the lateral view of the head alone to be a just representation of the fish described. Dr. Gunther, Catalogue of British Museum, vol. iii, p. 455, and Annals and Magazine of Natural History for May, 1861, p. 5, also, confines the name *M. chelo* to another species not hitherto recognised as British, the fish we have been accustomed to know by that name being his *M. septentrionalis*, the principal marks of difference between them being the decidedly shorter pectoral fin of the latter. In *M. chelo* this fin extends almost to the origin of the first dorsal fin. The upper lip is described as much thinner in the *M. septentrionalis*, the preorbital bones of a different form, and the tail more extended. It is certain, however, that these preorbital bones in our own fish, in their marginal teeth, as we represent them, are closely like those of *M. chelo* of Gunther, p. 454, and do not at all resemble those of *M. septentrionalis*, p. 455.

THIS species is less frequently seen than the last named, but when it appears, it is in far larger numbers and more huddled together. I have been informed of five thousand, and in another instance almost eight thousand that were taken at one haul of a sean. The usual season of success is in the winter or spring, when they enter harbours and appear busily engaged in searching the crevices of rocks and clumps of sea-weeds for their appropriate food. Their habits in other respects are but little known, except that they are disposed to seek their escape from confinement by leaping over an obstacle in the same manner as the Greater Grey Mullet. Although this species has only been distinguished from the other of late years, it is known, and even in considerable numbers, as far as to the extreme north of the United Kingdom.

I have not seen it larger than from ten inches to a foot

in length, and when many hundreds are taken together they usually appear to be of one size. Compared with the larger Mullet the body is less deep, the head of more uniform proportions, and the eye slightly higher on the cheek. First dorsal having four spinous rays, nearer to the second than its own breadth; second dorsal with nine rays; pectoral placed high on the side, with fifteen rays; anal ten, the last two from one root; ventral six, the first simple; caudal fin fifteen rays. The colour is much as in the larger Mullet.

Besides the species of Grey Mullets of which we have given figures and descriptions, there remain two or three others which are supposed to be natives of our coasts, and which therefore require attention; but I am compelled to acknowledge that I know nothing of them beyond what is related by Mr. Yarrell, and especially by Dr. Gunther, the notes of whom I shall content myself with transcribing for the use of such observers as may have the fortune to meet with specimens.

SHORT GREY MULLET.

Mugil curtus, YARRELL; Br. Fishes, vol. i, p. 244.

THIS supposed species is named and described by Mr. Yarrell from a single example obtained by himself while fishing at Poole, in Dorsetshire. The specimen scarcely exceeded two inches in length, and its principal distinction consisted in the extreme shortness of the body, which led to the adoption of the specific name.

The number of the fin rays was—of the first dorsal four, of the second one firm ray and eight others, pectoral eleven, ventral one firm and five others, of the anal three firm and eight others, of the tail fourteen rays. "The length of the head as compared with that of the body and tail, is as one to three, the proportion in the Common Grey Mullet being as one to four; the body is also deeper in proportion than in *M. capito,* being equal to the length of the head; the head is wider, the form of it more triangular, and also more pointed anteriorly; the eye larger in proportion; the fin rays longer, particularly those of the tail; the ventral fins placed nearer the pectorals, and a difference exists in the number of some of the fin rays. The colour of the two species are nearly alike; and in other respects, except those named, they do not differ materially."

Since the publication of the first notice of this species, the eleventh volume of the 'Histoire Naturelle des Poissons' contains a reference to this fish, of which an example was sent to Paris, and which M. Valenciennes considered to be identical with Mr. Yarrell's fish. It is to be regretted, however, that the specimen thus referred to by Mr. Yarrell is not to be found in the collection of that gentleman, the other portions of which generally were transferred to the collection of the British Museum. A fish must be exceedingly rare of which only two examples have been seen; but in the case of the Mullets the rarity becomes suspicious from the fact that it is the habit of all the species to keep near the shores. On the other hand a curtailment of proportions is far from being unusual in fishes, and when this takes place it commonly interferes with either the number or arrangement of other parts. On these accounts the Short Grey Mullet must remain for the present a doubtful species.

LONG-FINNED GREY MULLET.

GOLDEN MULLET.

Mugil auratus, RISSO. GUNTHER; Cat. Br. M., vol. iii, p. 442.

THE last-cited writer says that in the British Museum there are five specimens of this species obtained in England. Risso speaks of it familiarly as existing in the Mediterranean, and it is found in the Canary Islands; but with us it must be scarce, as the gorgeous colours ascribed to it by Risso would otherwise prevent it from being overlooked; and yet no observer on the coast has recognised it.

According to Dr. Gunther the height of the body is contained five times and one fourth to five times and three fourths in the total length, the length of the head five times; snout broad and depressed; cleft of the mouth more than twice as broad as deep; eyes with the rudiments of an adipose membrane; a short lanceolate portion of the chin not covered by the mandibular bones, (represented as much like that of our Lesser Grey Mullet.) Twenty-five scales between the snout and the spinous dorsal. No pointed scale in the axil of the pectoral fin. Risso says that the first dorsal fin has four rays, the second nine, pectorals seventeen, ventrals ten (of which the first is spinous,) caudal fin eighteen. On the gill-covers are some beautiful golden spots; the back bluish; on the sides seven well-marked lines; the belly of a brilliant silver; ventral fin reddish, anal white, tail light blue.

EIGHT-RAYED MULLET.

Mugil octoradiatus, GUNTHER; Cat. Br. M., vol. iii, p. 437, and Annals and Magazine of Natural History, May, 1861, p. 3.

THE proportions of the body of this species are much like those of the last named. It has eight soft rays in the anal fin; the eyelid not adipose; upper lip thin, but in the void space on the throat it differs from the Long-finned Mullet in having it lengthened out to a space between the interopercula, but it agrees with that fish in having twenty-five scales between the snout and the spinous dorsal. It differs from it again in the form of the lower jaw. The colour of the back greenish; sides and belly silvery, with darker stripes along the sides. Two examples only have been noticed, one of which was in Mr. Yarrell's collection, who mistook it for *M. capito.* It does not appear to have been known in any other country, or to have been noticed by any other observer.

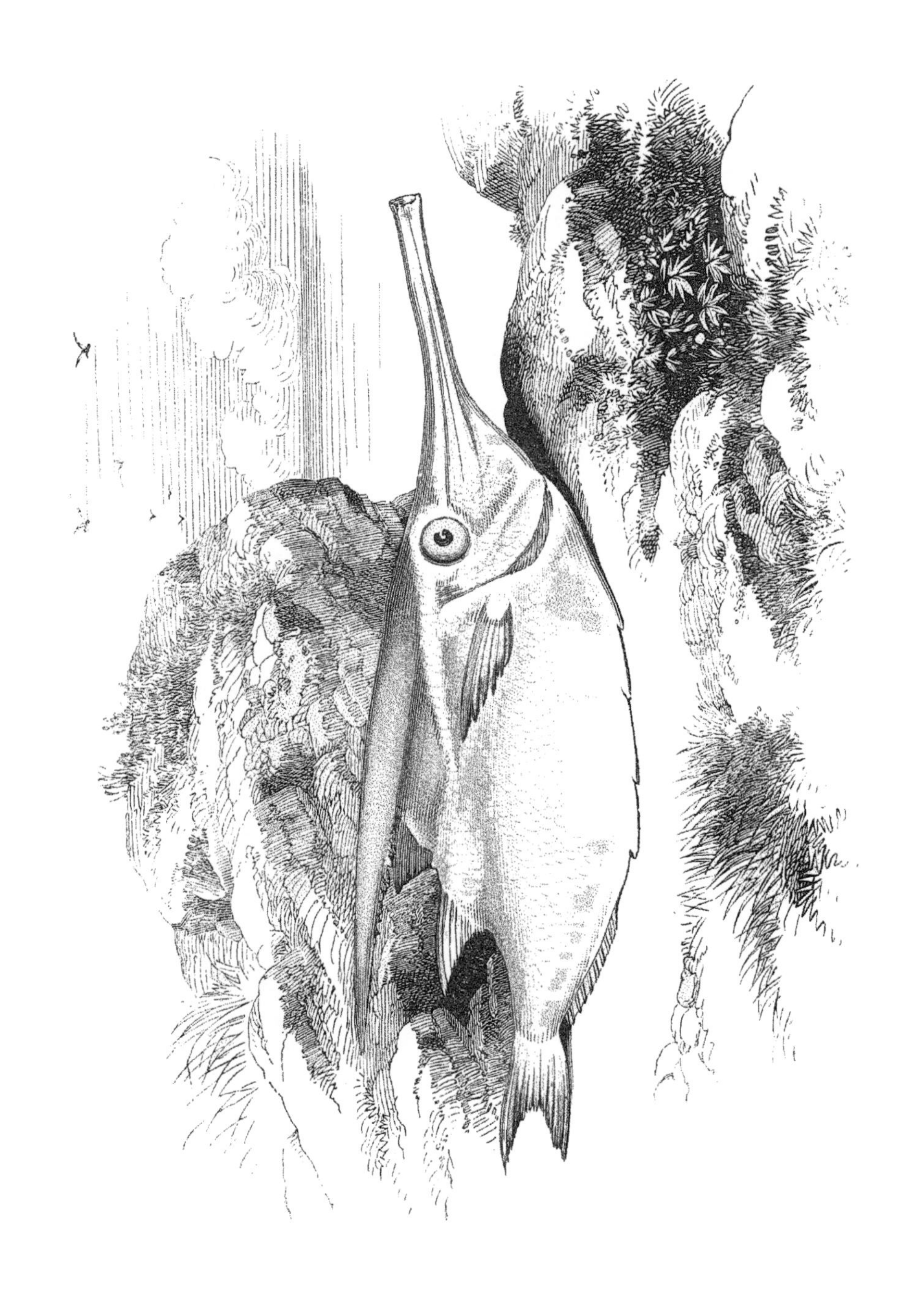

CENTRISCUS.

FORM of the body compressed, oblong, or elevated; the anterior bones of the skull brought forward into a long tube, at the end of which is a small mouth having no teeth. Body with a cuirass, or separate bony plates. Two dorsal fins, the first with a very strong spine; ventral fins small, on the belly.

TRUMPET-FISH.

BELLOWS-FISH.

Scolopax,	JONSTON; Table 1, f. 9, but this name is not derived from the same Greek word which signifies a woodcock, but from *Scolops,* a long and slender instrument, from which also the bird itself obtained its designation.
	WILLOUGHBY; p. 160, table 25, f. 2, representing a dry specimen.
Balistes scolopax,	ARTEDI AND LINNÆUS.
Centriscus scolopax,	CUVIER. BLOCH; pl. 123, f. 1, a poor figure.
Centrisque becasse,	LACEPEDE.
Solenostemus scolopax,	RISSO.
Centriscus scolopax,	FLEMING; Br. Animals, p. 220.
" "	DONOVAN; pl. 63.
" "	JENYNS; Manual, p. 400.
	YARRELL; Br. Fishes, vol. i, p. 346.
	GUNTHER; Cat. Br. M., vol. iii, p. 518.

THIS curious fish can scarcely be said to be common in any part of the Mediterranean, and it is scarce everywhere else. Willoughby met with some examples in the fish-market at Rome, whither they had been brought for sale as food; but at best, and in a country where very little birds and fishes furnish a supply for the table, these can add but little for that purpose, although it is admitted that what there is is of excellent quality. Risso speaks of it as not common about Nice; and Rafinesque, at Palermo, and Dr. Gulia, in his "Fauna of Malta," make no mention of it; which omissions may in part be explained by the information we obtain, that usually it is only procured after stormy weather.

The occurrence of the Trumpet-fish in Britain has only been in a few instances, of which two at least were in Cornwall. The first of these was thrown on shore in St. Austle Bay in the year 1804, and came into the hands of William Rashleigh, Esq., of the neighbouring mansion of Menabilly, who caused a drawing to be taken of it of the size of nature, and from which our own is a copy. It appears that Donovan had possessed two other British examples, from which he derived his figure, as above referred to; and the fragment of another was found on the beach in Mount's Bay in the year 1853, but it was too imperfect for preservation.

From the small mouth of this fish, with the absence of teeth, we may conclude that its food is the entromostraca, or minute animals of a variety of shapes that people the ocean as insects do the land; while its little aptitude for extensive motion will account for its limited wanderings, and consequently for its rare appearance in unaccustomed places.

The ordinary size of this fish is from four to five inches in length; and the following notes of other particulars are derived from a description made from the example taken in St. Austle Bay, as before referred to, at the time of its capture. "It was five inches long, and from back to the belly one inch and two eighths, in thickness three eighths of an inch; it weighed six drachms. It was red on the back, the colour becoming more faint on the sides, and the belly was silvery. The proboscis, which to the eye measured an inch and five eighths, was formed of a bony substance, which was continued along the back, where it terminated in a sharp point, spreading in the middle, where it makes an obtuse angle, just above a small fin behind the gills. The mouth, which is at the end of the proboscis, is covered with a valve that is fastened to the under part. The pectoral fin is small; it has two small dorsal fins, the former one having a very long spine, under which spine (and joined to it) are small projections like the teeth of a saw; there are three or four projections, very small, under the belly, which are hard, round, and transparent; the fins are whitish; the tail divided."

This description, imperfect as it is, might serve as sufficient to distinguish this fish from others, if it were not that we perceive from published figures that it is subject to some variety, and that a species much resembling it, but supposed to be distinct, has been discovered in the Mediterranean. The denticulations which we represent on the abdominal ridge—perhaps a little too strongly, if we may judge from the description given above—and of which Donovan and Mr. Yarrell take no notice, are a remarkable instance of this variation, as is also the form of the tail, which these two last-named observers represent as round, but which in our figure and the original description is described as divided. The other species that we have referred to, (*C. gracilis* of Lowe and Gunther,) and which has not only been found in the Mediterranean and Madeira, but even in Japan, is only different from *C. scolopax* in being somewhat longer in proportion to its depth, in having a much shorter dorsal spine, and conspicuously smaller scales.

For a fuller description of our own Trumpet-fish than is given above, we select the following notes from Willoughby, p. 160:—The body is covered with rough scales; snout very long, straight, narrow, growing wider towards the head; the mouth narrow and covered, in fact, by the small under jaw, the angle of which is depressed. Eyes large; belly with a sharp ridge; without ventral fins, which however are marked by two bones resembling teeth. A little behind this, on the middle of the belly, is a ridge having some small elevated teeth. Anal fin with eighteen rays. Dorsal fins two, placed far behind, the first being formed principally of one long and stout spine, which is capable of some motion upward and downward, but cannot be raised upright. On its under side is a channel, on each side of which is a row of teeth. In front of this larger spine is a very small one, and behind it three others. The second dorsal has twelve rays; the tail forked.

LABRUS.

The body oblong, and, with the gill-covers, covered with scales; lips fleshy; teeth prominent; a single dorsal fin, having two orders of rays, the first portion spinous, of which each one is tipped with a free membranous appendage. Ventral fins thoracic; the tail round or straight. The evenness of the border of the first gill-cover distinguishes this genus from *Crenilabrus*.

The name anciently applied to this class or family of fishes was *Turdus*, but, as a generic term this is now appropriated to the thrushes among birds; and in both instances, as well as in the English name of the latter, it holds the same meaning, the best known amongst them in each instance being mottled over with light-coloured spots. A disease of the mouth is also called thrush for the same reason. But the name of *Labrus* was also in use at a remote date, and is characteristic of their prominent and fleshy lips, which are the principal organs of acute sensation; but in English these fish bear the general name of Wrass, which is pronounced Wrath or Rath by fishermen of the West of England. The Rev. Mr. Johns, in his description of the Lizard Point in Cornwall, informs us that by the fishermen there they are called Raägh, which may be the ancient British term, as in pronunciation it approaches very near to the Welsh name Gwrach, which signifies an old woman; the Latin form of which, in the word *vetula*, we find to have been applied to more than one fish of a kindred shape. According to Rondeletius and Gesner, a name of the same signification, Vielle and Vieille, is applied to some of the same sorts of fishes in different parts of France.

BALLAN WRASS.

RATH. RAAGH. BERGLE, in the Orkneys. Pre-eminently the WRASS.

Turdus, JONSTON. WILLOUGHBY; p. 320, X. 1, but it is not easy to assign the proper synonyms of this species as of some others, because both these writers speak more of the colours of these fishes, which vary greatly in different examples, than of their distinguishing forms; as is the case also with Risso.

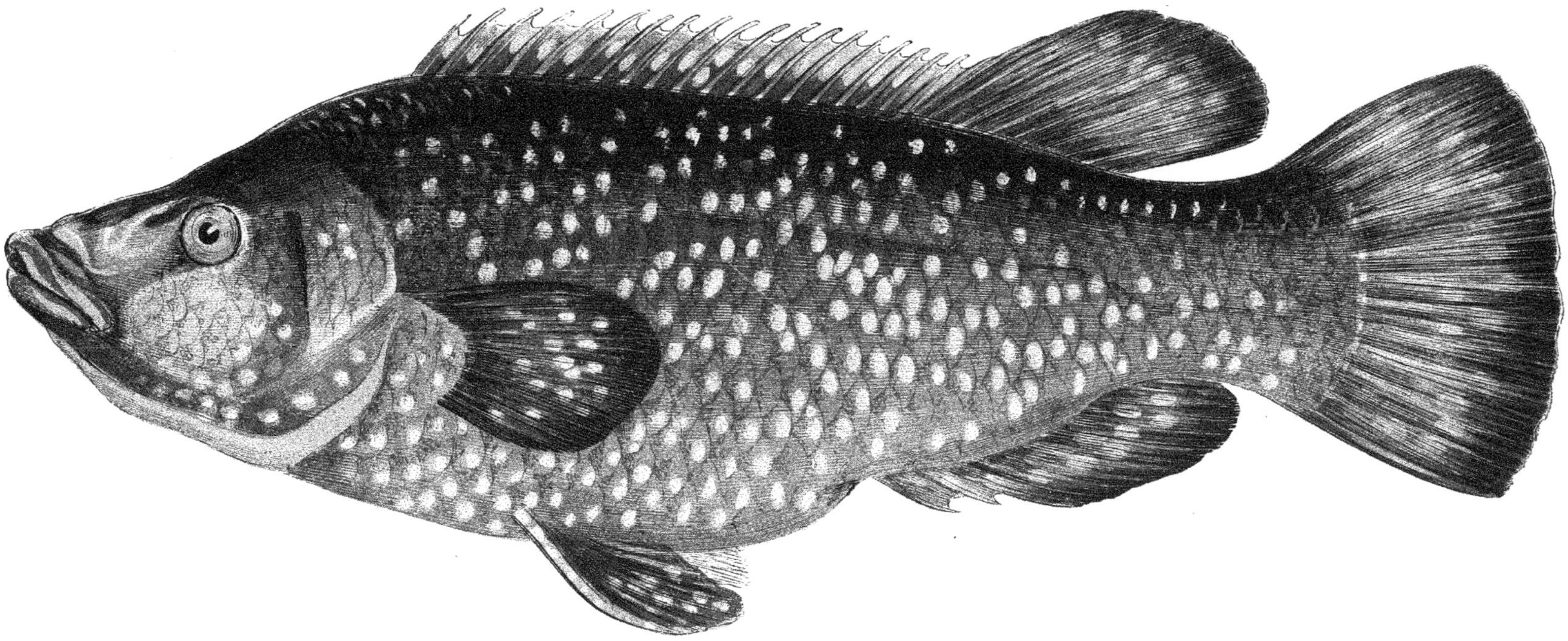

BALLAN WRASS.

CXXV

Labrus tinca, DONOVAN; pl. 83.
" *maculatus*, BLOCH. GUNTHER; Cat. Br. M., vol. iv. Dr. Gunther believes that the *L. pusillus* of Jenyns and Yarrell is the younger condition of this species.
" JENYNS; Manual, p. 391.
" YARRELL; British Fishes, vol. i, p. 311.

THE Wrasses form a numerous family, of which several species are found on the British coasts; and of these there are some portions of character which are common to all, however they may vary in other particulars. Their residence is among rocks which are clothed with the larger kinds of sea-weeds, and not at a great depth of water. If alarmed, or after wandering for a time in search of food, they return to their accustomed shelter; and as they appear to enjoy the waving of this herbage above their heads—among which they are seen passing to and fro, as if rejoicing in the pleasantness of the situation—they also find in the concealment it affords a shelter from their enemies, of which they have some formidable ones. The cormorant and shag can only prey upon the young; but the porpoise and dolphin, with the seal in more solitary places, will hunt them in the gullies of rocks, as I have seen the former do like hounds scenting out a hare, and from them there is small chance of escape.

The ancients entertained the singular opinion that the male Wrass was the master, keeper, or husband of several females, which he compelled to shut themselves up within some rocky prison, before which his occupation was to maintain over them a jealous watch lest any stranger should break in upon them to invade his rights. It was only in the evening that this care was intermitted, and only then for a short time, that he might seek a supply of food; the females, on the other hand, being supposed to find sufficient for their need in what their rocky cavern afforded. In no other way than by enticing this guardian of his flock to take the hook could a fisherman hope to ensnare the females within; but if successful in his first intention, his ultimate wishes were certain to be accomplished. It was then

"The females range unguarded by their mate,
Embrace the fraud, and share the common fate."
OPPIAN.

Observations such as these can only be supposed the sport of a lively imagination, but it is certain that in ancient times the fishes of this family were more observed, and held in higher estimation than now they are, and, indeed, as food they deserve to be. Pliny even calls Wrasses "*Turdi nobiles inter sexatiles*"—noble among fishes that frequent rocks; and Columella appears to give countenance to the same opinion by informing us that they were among the principal which the Romans kept in their salt-water stagna or ponds, and "quarem pretia vigent"—which sold for a good price; to which Oppian adds his suffrage by calling them delicate Wrasses. With us they are scarcely thought worthy of the trouble of conveying them to market, their flesh being considered soft and pulpy, without any distinguishing taste; but it is proper we should add, that although the Wrass is generally as little valued in Ireland as in England, we learn from Mr. Thompson, in his natural history of that country, it is otherwise in Galway; where a regular fishery for taking it is carried on, and it is preferred to most other sorts of fish. The truth, indeed, appears to lie between the two extremes, for when skilfully cooked it may maintain competition with some kinds that are held in better estimation.

Perhaps we may gather from Rondeletius some help that may assist us in explaining this variety of opinion in regard to the estimation in which this and some other fishes have been held as food at one time, and the dislike or neglect shewn to them at another. He remarks that the ancient Romans never ate their fish but with an artificial taste; so that their cooks were accustomed to shew their skill by dressing them with spices, a variety of herbs, and such strange sauces as we have already described. What would have tasted salt was rendered sweet with honey or sugar, and the insipid was highly seasoned. That which was tasteless was seasoned with onions, leeks, garlic, or omphacium, (a kind of oil or rob from the unripe olive, or the grape,) and vinegar; and these ingredients were perhaps necessary to render palatable what may have been long out of water in a warm climate. It is a remark sometimes made by our own fishermen that they could not eat what of this sort often finds access to a gentleman's table.

The Wrass is fished for from rocks overhanging the coast, or from boats near sunken rocks and gullies which they are known to frequent, the bait being a worm of the beach, or, what is to be preferred, a portion of some sort of crab, of which an example that has lately thrown off its covering or crust is the best. It is on such matters, together with several sorts of shell-fish and green sea-weeds, that they commonly feed. I have taken the limpet-shell (*Patella*) from the stomach. The bait is swallowed eagerly, but when hooked they struggle with much violence.

The beds of pharyngeal teeth, which are situated low down in the gullet, form a remarkable character in this fish, and require to be mentioned as being intimately connected with the nature of its food and its digestion. They are formed of two smaller triangular beds of blunt teeth above, with round tops, and planted on a bed of bone, and of a larger triangular bed opposed to them on the under side. And to render these teeth more firm for the work they have to perform, contrary to the example of teeth in the jaws of fishes, they are implanted in the substance of the bone itself, from which they appear to be renewed when their usefulness is destroyed. It is a question whether their employment is to act upon the food as it passes into the stomach, or rather that the grinding action is brought to bear upon it when it becomes regurgitated, in a manner which bears an analogy to the action of rumination in the ox and sheep. It is a confirmation of this last-named opinion that the sea-weeds and other matters usually existing in the stomach are found to be in a short time ground into a pulp. It might be supposed that the arrangement of these guttural teeth would lend assistance in the discrimination of doubtful species in this family; but observation shews that this is liable to some variation, and consequently must not be relied upon.

Fishermen have informed me that when they resort to a new station, it is usual at first to catch a Wrass—one or more—of the larger size, and afterwards, on going to the same spot, they find more in number, but of less weight; and from this they have drawn the conclusion that the older fishes had kept the young ones at a distance as long as they were able to retain the dominion. It may have been some incidents like this which led ancient observers to construct the story we have noticed above, of a tyrant male, from which the imprisoned females had been delivered by his death.

As the Wrass is not much sought after by professional fishermen, and for the most is only used as bait for other fish, or for lobsters and crabs, it sometimes lives long enough to shew signs of advanced age, the chief of which is partial or utter blindness; and in this condition they occasionally wander until their ill-chance leads them to the dangers of the shore. This defect of sight is sometimes produced by an opaque cloud, which covers the usual transparent cornea of the eye; and at other times it has its seat in the substance of the crystalline lens within, in which case it is the same with what in man is termed the cataract.

The Ballan Wrass is common on all the coasts of Britain, where the rocky bottom is such as to afford it food and shelter; and it is also found along the western shores of Sweden and Norway. It appears to be less common in the Mediterranean, and Risso says it is caught at Nice in July, as if it were subject to some periodical movement, which is not the case with us.

The spawn is shed in spring, and the young, of small size, are seen about the borders of rocks, at the ebb of tide, through the summer.

The Ballan Wrass is usually from fourteen to sixteen inches in length, with a weight of seven or eight pounds, and it has been known to reach the length of nearly two feet. The body solid, compressed, moderately deep, the shape sloping gradually from the nape to the point of the upper jaw, which protrudes a little beyond the lower. The lips fleshy and prominent, the upper more so than the lower, both having raised striated lines. Teeth firm, stout, slightly incurved, regular in the jaw; a double pair, more concealed, separate from each other, in front of the palate; a membranous veil forward in the mouth above and below, the latter occupying the place of the tongue. Jaws extensile. Nostrils above the line of the eye; a deep depression in front between the eyes. Eyes lateral, prominent. Body covered with oblong scales, the free portion of each clothed with a fine membrane in which the colour resides; the gill-covers also have scales, but none on the top of the head or before the eyes. Lateral line gently bent down opposite the termination of the dorsal fin. This fin begins a little behind the origin of the pectoral, with twenty firm rays, each one tipped with a soft process, the hindmost portion more expanded, having eleven soft rays; pectoral with fifteen; anal having two firm and tipped rays, and ten soft, the two last from one root; ventrals six rays, the first firm; caudal fin thirteen.

Colour lively, but very various in different individuals, the highest brilliancy very soon declining. Iris of the eye crimson, with a dark or purple border. The body yellowish, orange, or golden; back and top of the head brown; whitish or yellow, or mottled with orange on the belly; in some examples a general tendency to green, which is even to be discerned through the flesh. In the younger specimens there is often a beautiful and varied stripe of lighter colour, with touches of blue and pink, from behind the eye to the tail; sometimes blue spots on the tail fin. In the old individuals almost every scale is marked with a round spot of lighter colour, with a border of red, brown, or orange.

It is probable that in all the Wrasses the teeth are shed with regularity. They are hollow at the root, and, in the Corkwing especially, each one rises through its own membranous sheath to supply the place of another that is about to be thrown off. The depression referred to between the eyes forms a cavity that accommodates the retracted action of the complicated apparatus which is connected with the motions of the upper jaw, and which are guided by muscles that act through the means of tendons. A large muscle acts upon the angles of both jaws, to enable them to crush its food, while the curtains which lie across the mouth above and below are supplied with large nerves of sensation, derived from what may be called the facial nerve, and the lowest of the two branches being the largest. The whole structure of these parts points out the existence of a union of much strength with high sensibility of taste and feeling.

There is reason to believe that the Corkling of Jenyns and Yarrell, *Labrus pusillus,* is only a younger condition of the Ballan Wrass.

GREEN WRASS.

GREEN STREAKED WRASS

Labrus lineatus,	Donovan; pl. 74.
" "	Jenyns; Manual, p. 392.
" "	Yarrell; British Fishes, vol. i, p. 315.

We have seen, when speaking of the Ballan Wrass, that it is common to this whole family to be characterized by the possession of lively colours, which in each species are liable to considerable variation, and of which the intensity will be modified according to the nature of the ground they live in, or depth of water. But notwithstanding this tendency to vary, each species is found to possess a prevailing cast of colour, beyond a certain limit of which the variation does not proceed. These colours appear to have their seat in the epidermis or skin which clothes the body, and especially covers that elongated portion of each scale which remains free and not overlapped, and which serves to keep the scales in their place. Although the colour diffused over the body is intimately associated with their health and life, and even with their passions, so as to vary with these conditions in a very short time,—and the Ballan Wrass has been seen to change decidedly under the impulse of the fear of capture,—yet the prevailing bias of these tints appears to be under the dominion of chemical materials which are constituent portions of the blood, in the same manner as are the leaves of trees under similar conditions. Thus, in the Ballan Wrass, where the colours will be red, orange, or yellow, brown, blue, or green in different individuals, or on different parts of the same surface, yet gradually after death these colours will fade or change, and settle down into

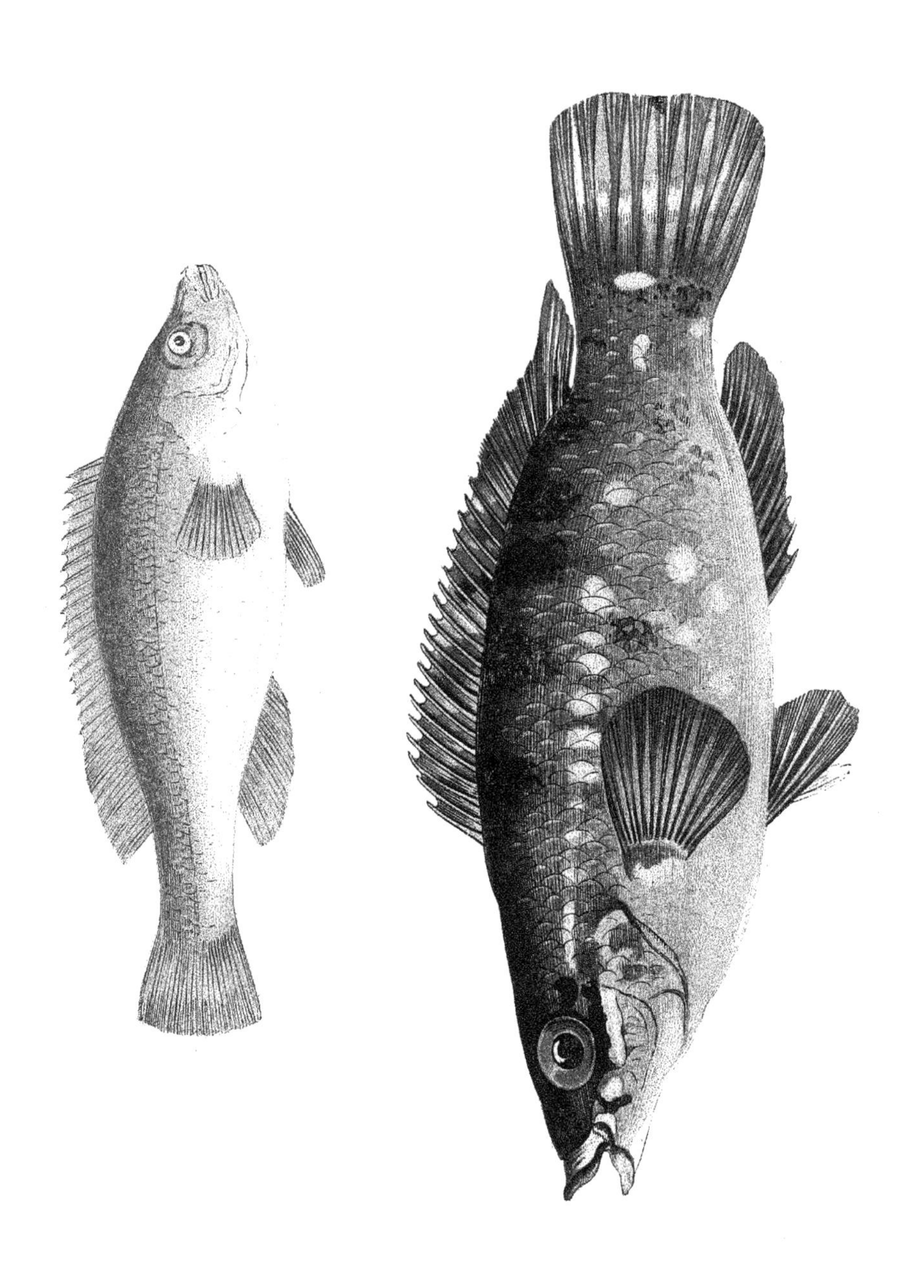

permanent green or red, as we believe according to the predominancy of their acid or alkaline affinities.

But there are species mentioned by foreign naturalists, which in life are said to be constantly marked with a preponderancy of living green; and as a fish similarly adorned is sometimes met with on our coast, observers have generally agreed to consider it a distinct species, with the name of the Green Wrass, or *Labrus lineatus*, the last-mentioned denomination being derived from some streaks of another colour that is seen upon it when of full size. But that the British Green Wrass is truly a separate species is far from certain, and our placing it under a separate name from the last species is rather in deference to the opinion of other writers than from our own judgment. That the situation to which they resort has much influence in producing the colour appears from the fact that those Wrasses which are found along that range of rocks on which the Eddystone lifts its light, and which consequently are several miles from land, are uniformly of a pale green, with some shades or lines of brown; but in other particulars, and especially of form, they are not to be distinguished from the common species. The younger fish are of the brighter green; and I am indebted for a figure of an example in this condition, of the natural size, as represented in the engraving, to William P. Cocks, Esq., of Falmouth, in which harbour it was caught by angling from the rocks. In this example the fin rays enumerated were,—in the dorsal twenty firm and ten soft, anal three firm and eight soft, pectoral fourteen, ventral eight, of which three were firm, in the caudal fifteen rays.

COMBER WRASS.

JAGO, in Ray's Synopsis Piscium, pl. 2, f. 5.

THIS fish must be distinguished from another Comber, *Serranus cabrilla*, in our "History of British Fishes," vol. i, p. 195. The only author besides Jago to which we can refer with confidence concerning it is Pennant, who has given a figure of an example he had obtained from Cornwall, and of which he says:—"It was of a slender form, the back, fins, and tail red, the belly yellow; the sides marked beneath the side line with a smooth even stripe, from the gills to the tail, of a silvery colour; the tail rounded at the end. The dorsal fin had twenty spiny and eleven soft rays, pectoral fourteen, ventral five, anal three spiny and soft, the caudal fourteen.

Dr. Gunther supposes that this fish may be the *Labrus Donovani* of Cuvier, which is described as having "the height of the body equal to the length of the head, and contained thrice and three fourths in the total; length of the snout one third of that of the head. Dorsal fin with twenty firm rays, and ten or eleven soft; anal twelve rays, of which three are firm." The colours are said by Valenciennes to be on the upper parts and fins green, a silvery band along the sides; head with some irregular blue lines.

This description scarcely leaves the subject more clear than before, but I will here introduce a fish to the naturalist, that if not the Comber Wrass, at least must be regarded as different from any other we have to describe as British. It was caught in a crab-pot, and measured in length five inches and a half. Head before the eyes lengthened and pointed, longer in proportion than in the Ballan Wrass; gape wide; lips very fleshy: upper jaw a little beyond the lower· teeth as in the Common Wrass; eye large; back rising to the dorsal fin; body compressed, moderately deep. Rays of the dorsal fin thirty, of which the soft portion has only four, which are suddenly lengthened, the first of these soft rays the longest; pectoral fin fourteen rays; anal three firm and nine soft; ventral six: caudal fourteen. The colour was beautiful, and with all the variety to which they are liable, such as I have never witnessed in the Ballan Wrass. The groundwork a rich deep mahogany red; on the cheeks a broad pale defined straight line backward from the border of the eye; between this and the angle of the mouth a broken pale line; all the lines bordered with a darker margin of the ground-colour; a similar narrow line, almost like a ridge, along the upper portion of the hindmost gill-cover. A more obscure wide interrupted line running back from the border of the gill-cover along the side of the body. Belly a little paler than the sides, approaching to reddish. Tail having broad transparent patches, six in all, somewhat regularly arranged, with dots of very dark brown at the base of the rays, and one larger and darker, but not exceeding the size of a large pin's head, at the end of the fin. Iris of the eye bright crimson; pectoral and ventral fins of the colour of the body. A few hours after death these spots generally vanished, and the colour became uniform. The drawing of this fish was made when it was alive in water, when it displayed marks of much activity. Another example with similar colours, but of considerably larger size, was afterwards caught.

In a communication with which I have been favoured from the west of Cornwall, the Comber Wrass is described as known to the fishermen of Mount's Bay, and as being "the most slender and most graceful of the Wrasses; the head smaller, lips thinner, jaws more lengthened and pointed:" but of this I have not examined a specimen.

COOK.

CUCKOO WRASS.

Blue-striped Wrass,	W. THOMPSON'S Ireland, vol. iv, p. 124.
Striped Wrass,	PENNANT. The Latin trivial names imposed on this fish are more in number than we shall refer to, but we know no reason why the original term applied to it by Jago, who first described it, should not be maintained.
Coquus, Cook,	JAGO; Ray's Synopsis Piscium, p. 163, f. 4.
Labrus variegatus,	DONOVAN; pl. 21, but too stout.
" "	JENYNS; Manual, p. 394.
" *coquus,*	YARRELL; Br. Fishes, vol. i, p. 317.
Labre melé,	RISSO.
Labrus mixtus,	LINNÆUS. GUNTHER; Catalogue Br. M., vol. iv.

THE Cook is one of the commonest of the Wrasses on the west coasts of the kingdom, and in its colours much the most brilliant; but it becomes more rare as we proceed northward, and is scarcely to be met with at the further extremity of the British Islands. I am indebted to the kindness of Mr. Iverach, of Kirkwall, in Orkney, for the information that he has known it caught once at that place, for I suppose this to be the species which he has mentioned in the belief that it was the *Labrus pavo* of Risso, a kind not yet known in the British Islands. According to Nilsson it is not uncommon on the coast of Sweden.

Its habits in the south and west of England are to perform a partial migration; or, rather, there is a change of quarters according to the season, as being found near the land when the weather grows warm, and through the summer and autumn it is not unfrequently caught by fishing from the shore. It

also enters crab-pots to feed on the bait, but it confines itself less to a limited range than the Ballan Wrass, and occasionally wanders about over level ground, from one tuft of sea-weed to another; and, in passing along, it sometimes meets with and swallows the baits of those who are fishing for pollacks in the manner termed whiffing. Its food is much the same as in others of this family, and I have taken fragments of a large galathæa from its stomach. It is rarely used as food.

It is seldom that this fish reaches to the length of a foot; the shape more lengthened than in the Ballan Wrass, and rounder; the snout protruded, conical, and fleshy; lips tumid and striated towards the mouth, with a fleshy overlapping; upper jaw a little the longest; gape moderate; teeth in a regular row, slightly hooked, none in front of the palate; veil in the mouth, above and below. Scales on the body and gill-covers, with only a slight border free; none on the head or before the eyes. Lateral line descending much more obliquely than in the Ballan Wrass. Eye lateral, rather large. The single dorsal fin begins above the root of the pectoral, with eighteen firm rays, tipped, and thirteen soft; the two last from one root; pectoral fifteen; anal fourteen, of which three are firm, and the two last from one root; ventral six, of which the first is firm; caudal fourteen.

The colours differ in different examples, but rather in distribution than tint, and they are always brilliant. In the specimen described, the eye was crimson with a purple border. Ochre yellow on the head, with pale purple stripes. Half the back brownish yellow, softening into orange red posteriorly, the belly pale red; all the fins more or less bright red; base of the dorsal anteriorly fine blue; border of the anal blue; high on the body broad purple lines running backward, and softening into pale blue; lines below more interwoven and broken; a broad dash of colour on the border of the tail above and below. In all the variations of colour, bright red, orange and yellow, with brilliant blue and purple prevail. The only spots are three or four of the size of shots, rarely seen on the membrane of the dorsal fin. The pharyngeal bones do not differ considerably, except in size, from those of the Ballan Wrass.

THREE-SPOTTED WRASS.

Trimaculated Wrass,	Donovan; pl. 49.
Labrus trimaculatus,	Turton's Linnæus. Cuvier.
" *carneus,*	Bloch.
Labre tripletache,	Risso.
Labrus trimaculatus,	Jenyns; Manual, p. 396.
" "	Yarrell; Br. Fishes, vol i, p. 320.
" "	Gunther; Cat. Br. M., vol. iv.

This is also a common fish, as well in the Mediterranean as on the west and south coasts of England and Ireland; it is also mentioned by Nilsson as met with in Sweden; but it is rare in Scotland, and is only of casual occurrence in the Orkney Islands.

Its habits and food resemble those of the Cook; and indeed it is the opinion of Nilsson and Dr. Gunther that it is the female of the last-named species; a fact which future observation must decide. It spawns in April or May, and on the second day of the last-mentioned month an individual was examined that proved to be a sharer of both sexes. The mature roe passed from it on slight pressure, but, on cutting the body open, while one lobe was found nearly empty, the other was far short of perfection, and a lobe of milt, in the same condition, lay with them.

This fish reaches the length of eight or nine inches; the shape rather lengthened, plump, moderately compressed, not so robust as in the Cook, which in outline it generally resembles. Head lengthened before the eyes; jaws equal; lips fleshy; teeth in the upper jaw numerous, those in the front large, separate, curved, projecting; in the lower jaw the two corner teeth in front like the upper front teeth, the others small. Body covered with scales as in the others of this family; lateral line descending gradually opposite the termination of the dorsal fin.

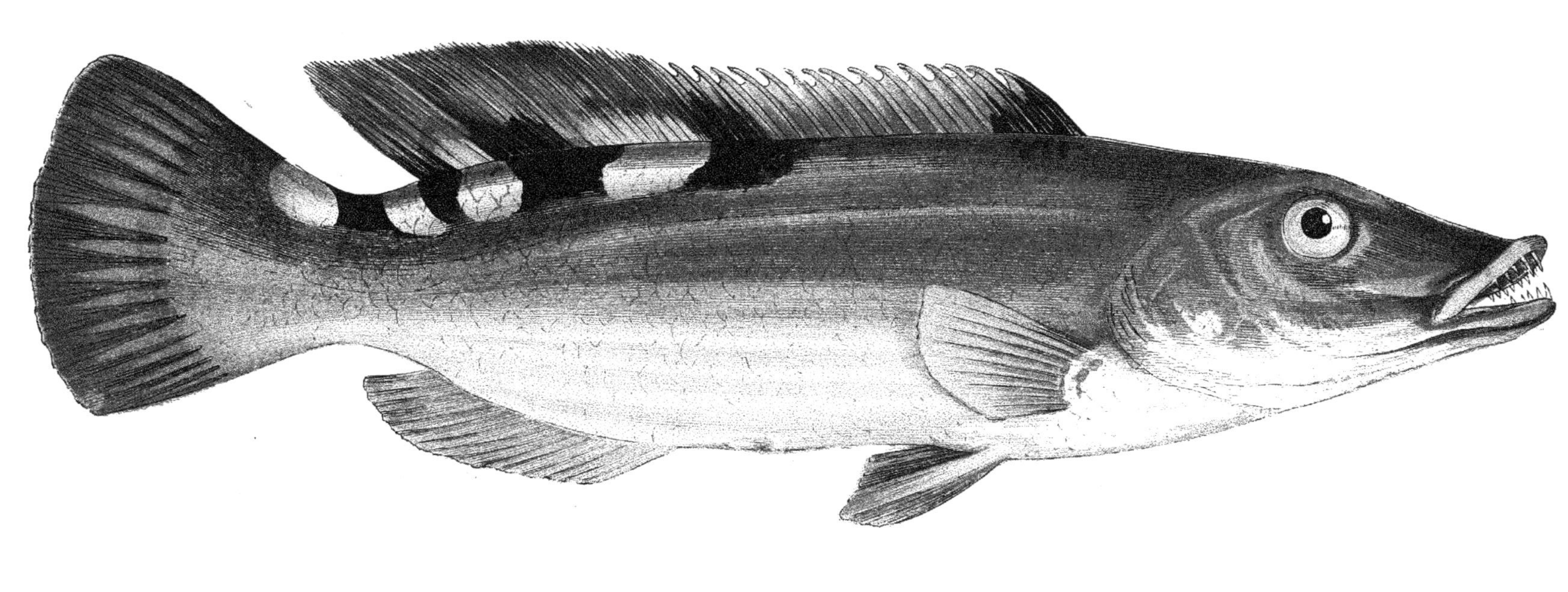

THREE-SPOTTED WRASS.

CXXVIII

The dorsal fin begins above the base of the pectoral, with thirty rays, of which sixteen are firm; anal fin has two firm and twelve soft rays: both these fins expanded posteriorly, and end opposite each other. Tail round, nineteen rays; ventrals six. There is some variety in the colour, but chiefly as regards the depth or intensity, those examples which live in the deeper water and most rocky ground being generally of a lively vermilion, paler on the belly.

This fish obtains its name from some remarkable very dark spots on the back and dorsal fin, in many cases amounting to five, and I have not known them less than three. One is at the origin of the dorsal fin, and is sometimes wanting; the others are at the hindward part of the back, extending up on the dorsal fin, except the last, which is on the back behind the termination of that fin; and alternate with these mentioned spots are an equal number of pale or flesh-coloured spots, ending with the latter. I have known an example where the red colour was more dull, with the absence of the spot at the origin of the dorsal fin, and of all the light-coloured spots. The tail also had a slight border of blue.

SEAWIFE.

Acantholabrus Yarrellii, YARRELL, British Fishes, vol. i, p. 339.

THE only example on record of this supposed species was not preserved, and Professor Nilsson believes that it was an irregularly-formed individual of the Cook; to which opinion Dr. Gunther agrees—"Catalogue of British Museum," vol. iv.

ACANTHOLABRUS.

The body oblong, moderately compressed; dorsal spines many, as of twenty or more; anal spines more than three.

The last particular in this generic definition is the only one in which it differs from the other sections of the family of Wrasses; and, however convenient it may be for naturalists to constitute distinctions for the arrangement of a rather numerous family, it must be remembered, that so apparently unimportant a matter, in itself also liable to variation, is not sufficient to constitute a natural separation of species into genera.

SCALE-RAYED WRASS.

Labrus luscus,	Loudon's Magazine of Natural History, vol. v.
" "	Jenyns; Manual, p. 400.
Acantholabrus Couchii,	Cuvier.
" "	Yarrell; Br. Fishes, vol. i. p. 336.
" "	Gunther; Cat. Br. Museum, vol. iv, p. 92.

The earliest account of this species, (after the short note, with a sketch, in Loudon's Magazine, as above referred to,) is in Mr. Yarrell's "History of British Fishes;" and the scarcity of this fish may be supposed, when no other had been met with for a very long time. It happened, however, that as a fisherman lay at anchor off the Deadman Point, on the south coast of Cornwall, where the depth of water was above fifty fathoms, an individual of this kind of Wrass took his hook. He was about to cut it in pieces, as is the common fate of the Wrasses when made bait for other fishes, and had already cut off a slice from the side, when it appeared to him that it was of a sort he had not seen before: it was accordingly preserved for my use. A drawing in consequence, as well as a description, was taken from it; and after being preserved.

as well as it could be done under the circumstances, by Mr. William Laughrin, A.L.S., of Polperro, the specimen was handed over to the collection of the British Museum.

The habits of this species are of course but little known; but although probably local, it may, within its range, be as common as others of its family. It appears, however, that their resort is in deeper water than such as most species of Wrasses prefer; and if the Scale-rayed Wrass has been caught by fishermen who often seek their livelihood in its favourite districts, it is not likely that in general they would otherwise notice it than as something fitted to their individual use. On many occasions has the information reached me of the capture of fishes which, from the attention they excited, were evidently of uncommon occurrence, but which have afterwards been thrown aside, although at the same time a little reflection would have called to the remembrance of these poor men that the presentation of a rarity was certain to meet its reward.

As some doubt has been thrown on the synonyms of this species, I regard it of importance to copy the notes which were made from the first specimen at the time of its capture, to which I will add the further notes that were obtained from an inspection of the second example above referred to. Of the first it is said, "The specimen was twenty-two inches long, the greatest depth, exclusive of the fins, two inches and a fourth; the body plump and rounded. Head lengthened; lips membranous; teeth numerous, in several rows, those in front larger and more prominent, slightly incurved. Eye moderately large. Anterior plate of the gill-cover serrated; body and gill-covers with large scales. Lateral line nearer the back, descending with a sweep opposite the termination of the dorsal fin, and thence backward straight. Dorsal fin with twenty-one firm and eight soft rays; the hindmost portion of the fin expanded; pectorals round, fourteen rays; ventrals six, one of them simple, stout and firm: between these fins a large scale. Anal fin six firm and eight soft rays, the soft portion expanded; tail round, fifteen rays. Between each of the rays of the dorsal, anal, and caudal fins is a process formed of firm elongated imbricated scales. Colour uniform light brown, lighter on the belly; upper eyelid black; at the edge of the base of the caudal fin a dark spot; pectorals yellow; dorsal bordered with yellow."

"I have never seen more than one specimen of this species, which was taken with a line in February, 1830, at the conclusion of a very cold season. It differs from the Common Wrass and Corkwing in its more elongated form and rounder make; from the former also in having a serrated gill-cover. From the Rock Cook it may be readily distinguished, besides its greater size by its longer form, larger mouth, rounder tail, and by the spot at the root of the tail being further back. The eye is larger than in the Cook, and nearer the angle of the mouth. It may also be distinguished by the serrated gill-cover, form of the lateral line, by the more numerous tipped dorsal and anal spines, and less rounded tail, as well as by the colours, which are sober, whereas those of the Cook are always splendid, and are little subject to variety, except of distribution. From all the species with which I am acquainted it may readily be known by the singular imbricated process of scales between the rays of the fins as above described."

Of the second example, (the lower figure on the plate,) I remarked at the time of receiving it that it had much the appearance of a Serranus. It was ten inches long, and more compressed than the former; flat on the head, and rising from behind the eyes to the root of the dorsal fin. But in colour it was paler than the former example, and without its finer marks; dusky pink over the body, a dash of dark over the eye; the eye itself silvery, with a blue border. A black spot on the dorsal fin where the different kinds of rays join, but not colouring the rays themselves. A broad spot on the upper margin of the root of the tail, and another fainter and more scattered on the lower border. The dorsal fin twenty-nine rays, of which twenty-one were firm; anal twelve, of which five were firm, being one firm ray less than in the former specimen; pectoral fifteen; ventral six, of which the first is firm; the caudal fin had the stems of the rays so covered with scales that they were not counted. I could not discover a nostril.

Dr. Gunther considers this last specimen as an example of *Acantholabrus Palloni* of Cuvier, *Lutjanus Palloni* of Risso, a species not otherwise known in British natural history.

ROCK COOK.

SMALL-MOUTHED WRASS.

Labrus exoletus,	LINNÆUS.
Crenilabrus microstoma,	COUCH AND THOMPSON; a paper by the latter, with a coloured plate, in the second volume of the Magazine of Zoology.
Acantholabrus exoletus,	CUVIER. YARRELL; Br. Fishes, vol. i, p. 341.
Centrolabrus exoletus,	GUNTHER; Cat. Br. Museum, vol. iv, p. 92.

THIS little fish is not rare on the western coasts of the kingdom, and in Ireland; and if it appears to be so it proceeds from its habits, which again are influenced by the small size of its mouth, that prevents it from taking the hooks usually employed in fishing at the stations where it is commonly found; for it seems ready to devour its food, and it is sometimes found in crab-pots, which it has entered to feed on the baits suspended there. As the crab fishery is not carried on in the stormy season of the year, this fish is for the most part only met with in the months of summer. Nor does it wander into such shallow water as some others of the smaller species of this family; yet it seems to be widely spread, at least towards the north, for Nilsson mentions it as inhabiting the coast of Sweden, where, however, it is not of frequent occurrence.

It does not exceed four or five inches in length, and one of the former dimensions was in depth one inch and four tenths, with plump appearance, although moderately compressed at the sides. The gape is small; lips fleshy; teeth regular and blunt; a veil or membrane at the palate, as in the other species; but in this it forms an arch on each side. Eyes rather large; head flat above, and in many examples it is elevated prominently on the top of the head; first plate of the gill-covers minutely serrated; a row of prominent mucous orifices round the posterior half of the eye. Cheeks and body covered with large scales. The lateral line first rises, and near the tail drops again. Dorsal fin with seventeen firm and eight soft rays, the last double, with a slight sulcus under it; anal fin with five firm and seven soft rays, the last also double; ventrals six rays, of which one is firm, the fin partly fastened down with a membrane. Pectoral fin round, thirteen rays; tail nearly straight. Scales on the roots of all the fins except the pectorals. Colour of the back dusky red, brighter on the head and lips; hindmost gill-covers red; a slight gold-coloured line from the eye backward; pale green on the belly; near the upper margin of the tail, between it and the termination of the dorsal fin a broad and conspicuous dark mark. Fins inclined to red, the three first rays of the dorsal bluish black. Eye crimson.

As in the other species of this family, there is in this a disposition to vary in the distribution and intensity of the colours, and even in the form, the elevation of the top of the head being much more considerable in some than in others. In rare instances the broad spot near the tail is wanting. Again there are in some examples longitudinal bands of colour on the dorsal fin, and curved lines of bluish purple on the cheek.

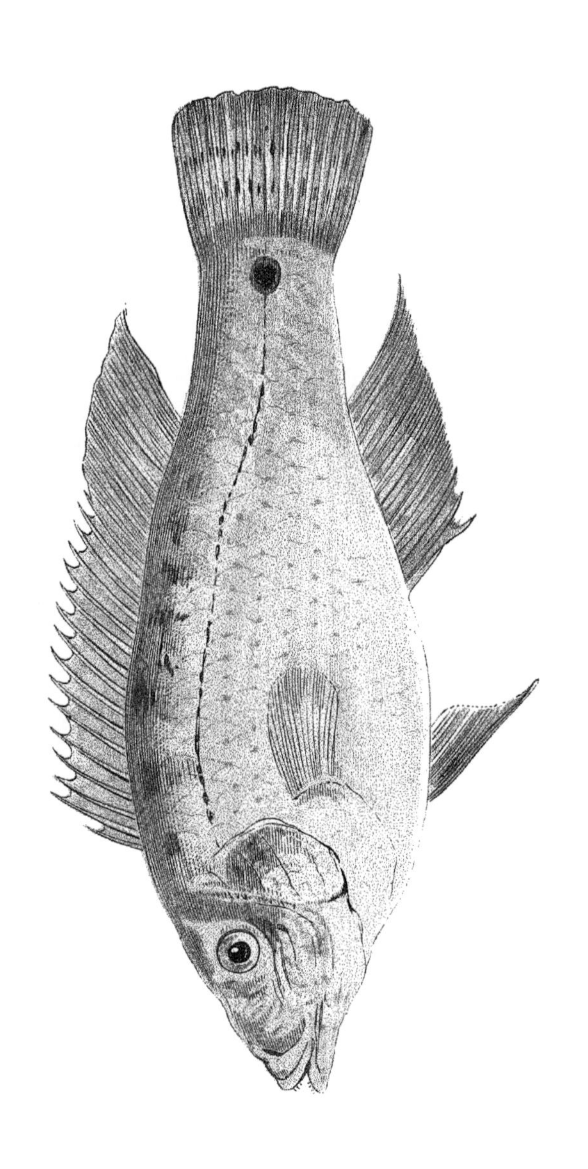

CORKWING.

Goldfinny,	PENNANT; pl. 97, 3. DONOVAN; pl. 72.
Labrus cornubiensis,	TURTON'S Linnæus.
Crenilabrus cornubicus,	YARRELL; British Fishes, vol. i, p. 328.
" "	JENYNS; Manual, p. 398.
" *melops,*	GUNTHER; Cat. Br. M., vol. iv, p. 80.

THIS is the most abundant of the Wrasses which are found on the west coasts of the kingdom, so that a boy has been known to catch sixty with a line at one time. It is perhaps less numerous as we proceed northward, but it is met with on the coasts of Sweden and Norway. It keeps in shallower water than the generality of this family, and is often seen in the gullies of rocks half uncovered by the ebbing tide, where it feeds on crustaceous animals, and is always ready to take a bait. In search of food it also wanders in various directions; and an observer sitting on a rock as the tide is flowing may perceive considerable numbers passing hastily in companies of two or three from one clump of oreweed to another, along the beach, for it is in such situations they expect to supply their wants, and they do not remain long at a distance from concealment or shelter.

In common with other Wrasses, the roe of the Corkwing is shed in spring; and the young, of a pale green colour, and perhaps less than an inch in length, are seen basking or loitering in shallow water on the borders of rocks through the summer. The full-grown fishes will also remain in pools of the rocks, where they are not unfrequently caught in the shrimp-net. No use is made of this fish as food.

The Corkwing seldom exceeds five or six inches in length; deep in the body, and compressed, so that in many instances the depth will be one third of the length. The head slopes in more or less of a waved line from the origin of the dorsal fin to the snout; jaws almost equal; gape narrow; teeth in regular order, but often in irregular stages of development; first plate of the gill-covers serrated. Large scales on the cheeks and body. Lateral line nearer the back, bending down suddenly opposite the termination of the dorsal fin. This fin is furnished with sixteen firm and nine soft rays; pectoral round, fourteen rays; ventrals close together, with one firm and five soft rays; anal fin three firm and nine soft rays, the two last from one root; tail fourteen rays.

Like others of this family, the Corkwing varies in its colour. Upper part of the head and back usually brownish, with stripes of red and green on the gill-covers; sides a faint green, with numerous lightish red stripes, yellow or greenish below. The fins are similarly varied, but the colours are commonly fainter than in others of the Wrasses. Near the tail, close to or on the termination of the lateral line, is a conspicuous black spot, which is seen in the earliest stage of growth of this fish, and its absence forms the only distinguishing mark of the supposed species, called the Gibbous Wrass by Pennant. This last-mentioned fish received its name on account of a considerable elevation of its back, and consequent greater depth than exists in others of this family, with a greater sweep downwards to the mouth; but the almost unanimous opinion of observers has concluded that it constitutes only an older and perhaps better fed stage of the Common Corkwing, from which the lateral spot near the tail has disappeared. It has been met with of the length of nine inches, with a deeper profile than is usual with the Corkwing; but I have never seen the back so high as is represented in Pennant's figure. Those also that have fallen under my notice have come from a greater depth of water than is usual with the Corkwing.

BAILLON'S WRASS.

BAILLON'S WRASS.

Crenilabrus Baillonii, GUNTHER; Cat. Br. Museum, vol. iv, p. 81.

THIS species is now for the first time introduced among British fishes, and that too with some degree of hesitation; but a drawing of one which came a few years since into my possession, and which then appeared to differ from the ordinary appearance of the Corkwing, and especially from that variety of it known as the Gibbous Wrass, conveys so near a likeness to a species described by Dr. Gunther, in the "Catalogue of the British Museum," above referred to, that I feel little doubt of its being the same fish. To provide the requisite evidence of this, and to render our account of it the more complete, the observations of Dr. Gunther are here added to our own description; but it is necessary to remark, that whereas considerable stress is laid in these extracts from the catalogue on the black marks described as existing on the dorsal and anal fins, the further description distinctly shews that, as well this species as the other which is said to bear a close affinity to it, are liable to much variety in this respect, at least as regards their distribution.

Our example measured seven inches and a quarter in length, and two inches in depth; covered with large scales, as well the body as the gill-covers; a border of small scales, prominently marked, round the posterior half of the eye. The profile descends from the origin of the dorsal fin to the mouth; jaws equal. The proportions generally as in the Corkwing, but the dorsal and anal fins are carried nearer to the root of the tail. Pectoral and ventral fins dusky yellow, the former with dusky bars; general colour of the other fins, head, and back very dark brown; on the sides longitudinal streaks from the pectoral fins backward, the whole, including the root of the pectoral fin, mottled with streaks and spots of very dark brown; tail with cross bars, and, near the termination of the lateral line, an ocellated spot.

Dr. Gunther's account of *Crenilabrus Baillonii* is thus given:—"The height of the body is contained three times and a third, or three times and a half in the total length. Cheek with two or three series of scales; the length of the snout is one third of that of the head. The soft dorsal and the anal with two black or blackish spots at the base. Back with five or six dark cross bars, more distinct in young individuals. Operculum without dark spot; base of the pectoral black. The dorsal fin with fourteen firm and ten soft rays; anal three firm and ten soft rays."

"This species will be easily distinguished from its nearest ally, *C. quinquemaculatus*, by the larger scales on the cheek, and by the greater number of longitudinal series of scales above the lateral line. The characters by which the two species have been distinguished are constant in all the specimens examined. The anterior spot on the dorsal fin is the most distinct, whilst the others are paler, and may entirely disappear with age. The number of the anal rays has been incorrectly given by Valenciennes. This species has been found on the coasts of Lisbon and Mogader, as also in the British Channel."

The *Crenilabrus quinquemaculatus*, here referred to, is a native of the Mediterranean, but has not been recognised as an inhabitant of our own shores. It is subject to much variety, and sometimes has a dark spot on the middle of the base of the caudal fin, (as in the fish we have described; but in our example it was ocellated, or marked with a pale border, as we have never seen in the Corkwing.)

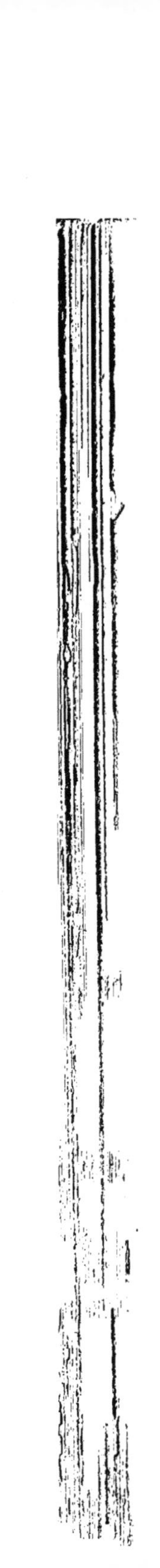

JAGO'S GOLDSINNY.

Goldsinny,	JAGO; in Ray's Synopsis Piscium, f. 3.
"	YARRELL; British Fishes, vol. i, p. 333.
"	THOMPSON; Nat. Hist. Ireland, vol. iv, p. 129.
Ctenolabrus rupestris,	CUVIER. GUNTHER; Cat. Br. M., vol. iv, p. 89.

IN the west of England this is a common species, and in many districts besides it is so also, although the individuals are not in considerable numbers. It occurs along the coasts of Scotland, as well as in the Baltic. Nilsson reckons it among the fishes of Sweden, and on the other hand it is met with in the Mediterranean. Unlike the Corkwing, I have not found it to frequent tidal harbours, and it appears to prefer deeper water than the shallows of the sea afford. It takes a bait, but is most frequently caught in the crab-pots, which it enters for the purpose of nibbling the baits suspended within them. It is probable it sheds its roe in May, as the Rock Cook is also known to do.

Its usual length is about five inches, with a depth of an inch and a quarter at the ventral fins; from the front of the dorsal fin the outline slopes gradually to the upper lip; jaws equal; teeth prominent; eye rather large. The body covered with large scales; gill-covers with scales of smaller size; lateral line suddenly bent down a little behind the termination of the dorsal fin: of this fin the rays are twenty-six, of which seventeen are firm and tipped; anal fin with three firm rays. The colour of this species is confessedly prone to vary, while the spots are believed to constitute a constant character. In the example described the upper portion was reddish brown; cheeks red, yellowish posteriorly, and a deep pink lengthened spot at some distance below the eye, which is likely to be accidental; the borders of the scales tinged with pink, which on the sides give the appearance of faint pink lines; pectoral, dorsal, and caudal fins inclined to red. The front of the dorsal bears a dark mark, which extends to three or four of the rays; at the upper border of the base of the tail is a round or oval spot, which may rather be said to be on each side of the edge than exactly on the top; a pink spot close to the upper part of the base of the pectoral.

This species of Wrass remained long in a state of uncertainty, from having been confounded with such others of the *Labridæ* as had their chief marks of distinction in a spot at or near the root of the tail. Mr. Yarrell led the way in part to the detection of this error, by engraving, although only as a tail-piece, a figure of the fish at p. 301, vol. i, of his first edition; but the likeness was much distorted by the accidental pressure of the specimen in its conveyance. Further doubt, however, appeared to be dispelled by Mr. Selby, who published in the first volume of the "Magazine of Zoology," etc., p. 167, with a plate, a description of the fish from examples thrown on shore in a storm on the coast of Northumberland. But the figure given by Mr. Selby is more stout and deep than any of the examples I have seen; and a remark to the same purport is made by Mr. M'Calla, as quoted in Mr. Thompson's work, already referred to. If the form of the scales in Mr. Yarrell's plate is correctly represented, there is in these also a remarkable difference in these fishes; and the broad bands from the back downward, as in Mr. Selby's figure, have never presented themselves in a Cornish, nor, I may add, in an Irish example. On the contrary, I have had occasion to mention the pink-tinted horizontal lines on the sides, which have also been noticed in Irish specimens, but not in those of the north of England. How far these differences may be explained by reference to the variety of ground and climate in which the individuals are found, is a subject for further inquiry; and this observation is the more appropriate, as, according to Dr. Gunther, who quotes the work on Ichthyology by Cuvier, there is a species with which it might be confounded. This is the *Ctenolabrus marginatus*, which is a native of the Mediterranean, and of a more lengthened form than this *C. rupestris*, with the same number of fin rays, and a large black spot anteriorly on the dorsal fin, and another on each side of the caudal fin; the vertical fins with a narrow blackish margin

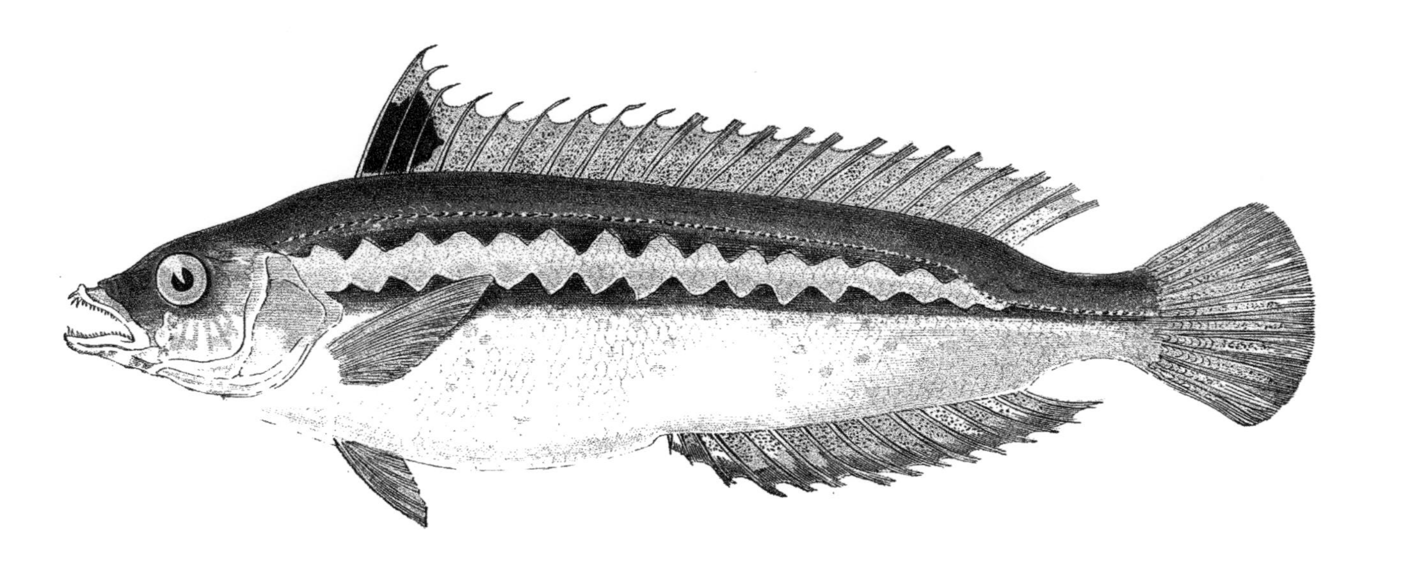

RAINBOW WRASS.

CXXXIV

JULIS.

This genus differs from others of the family of Wrasses in having the head entirely without scales. The lateral line forms an angle opposite the end of the dorsal fin; to which Swainson adds, that the first rays of the dorsal fin are higher than the succeeding, and the fin itself is thus rendered falcate.

RAINBOW WRASS.

Julis,	Jonston; pl. 14, f. 3.
"	Willoughby; pl. X. 4, p. 324.
Labrus julis,	Linnæus. Donovan; pl. 96.
Julis vulgaris,	Cuvier. Fleming; Br. An., p. 210.
" "	Yarrell; Br. Fishes, vol. i. p. 344.
Coris julis,	Gunther; Cat. Br. M., vol. iv, p. 195.

So far as is known this very pretty fish has only occurred in a single instance in England, which was in Mount's Bay, in the county of Cornwall; which district may therefore be regarded as the furthest extent of its range northward. It appears to have been caught by the merest accident, in the year 1802, and fortunately was obtained for the use of Mr. Donovan, who has given a beautiful likeness of it. But its history is to be learnt from the Mediterranean, where, in some districts, it is common, and where from ancient times some strange stories are told of its habits. Oppian relates in verse what Ælian repeats in prose. The last-named writer says that this fish lives in rocky places, and has its mouth poisonous, in such a manner that whoever tastes it is rendered unable to swallow. When fishermen have caught a Prawn, (*Squilla,*) the middle portion of which has been devoured by a Julis, they sometimes eat it; but the consequence is that they experience severe pains in the bowels. Swimmers and divers are much molested by this fish, in the same manner as flesh-flies assail and bite them; and Oppian compares the effect to the sting of a nettle. These divers are compelled to drive the fishes away, to avoid being tormented with their bite; and so persevering is the annoyance, that the men are obliged for the time to give up following their occupations.

However foreign to truth this account, and especially the first portion of it, may appear to us, we should call to mind, in vindication of the writer, that he reports no more than the current opinion of his day, and that the particulars themselves were in close connection with the theories then predominant. It was long held as a principle in natural philosophy, that the sea contained something in every instance that bore an analogy to what was in the sky above, or on the land; and the attention of the philosopher was directed to the discovery of such objects as were thus believed to carry out these correspondences of nature. Many figures, with a little violence done to the likeness, are in this light handed down to us by writers of the middle ages, who had not yet escaped from the trammels thrown around them by the ancients; and it was with these impressions that they imposed a name upon a fish because they supposed it to be endued with some of the ill qualities which belonged to an insect with which they were acquainted. The Creeping Julus was said to convey a poisonous bite; and that the fish Julis will annoy and bite we have the authority of no less an observer than Rondeletius, who describes what happened to himself, as well as on another occasion to a friend. When on one occasion he went to bathe near Antipolis, he saw several of these small fishes hasting towards him, and they attacked with their bites not only his legs, but the hard portions of his heels; and a similar circumstance was related to him by some gentlemen, as happening to them near Nice. No injurious effect followed to this eminent naturalist; but it is highly probable that the terror which would arise to an ordinary person from the prospect of danger, would confirm the impression that the danger itself was not wholly imaginary. There is reason to believe that these fish are usually in companies. Their food and season of spawning are for the most part the same as in other kinds of Wrasses; but they are little regarded as food, those which keep in the deeper water being considered the best.

The usual size of the Rainbow Wrass is in length from four to six or seven inches; the shape round and slender, so that Willoughby, from whom much of our description is derived, compares it to a Goby or Blenny. The mouth small, pointed; lips fleshy; teeth in one row; those in front larger and longer, especially a pair in the upper jaw. Eyes small. Head without scales; body covered with them; lateral line bent angularly opposite the termination of the dorsal fin. The dorsal fin high at the beginning, the first rays close to the head, more slender in its progress, reaching near the tail, having twenty-one rays, of which nine are firm; anal fin two firm and twelve soft rays; pectoral fourteen; ventral six, of which one is firm; tail round, twelve rays.

The colour is subject to some variation, but is always beautiful; and the males are said to excel the females in this respect. Along the back it is dark, (Risso says bluish green;) the belly blue or whitish. From the snout a variously-coloured line runs through the eyes to the middle of the sides; that portion which is on the cheeks saffron-coloured, passing into black, and in its further progress blue. Along the side from the gill-covers to the tail a wide line, with an irregular border of a light blue, and parallel to it below, a line of bright yellow. Eyes red or yellow. Dorsal fin with a band of yellow near the back, followed above by red, the upper border blue; on its front a distinct pink mark, which extends to the three first rays, and above this a black spot, including the second and third ray; tail yellow with a cast of red; anal fin coloured like the dorsal.

GADIDÆ.

THE CODFISH KIND.

The body moderately lengthened, compressed, with numerous small and soft scales; none on the head; more than one fin on the back, all with soft rays. Teeth in the jaws and front of the palate; the tail separate from the other fins, straight or round. Ventral fins under the pectorals. Thoracic fishes.

Linnæus classed the whole of this family in one genus, but the species have since been separated by Cuvier into several genera, according to organs which have an influence on their manner of life. It is to be regretted that the original Linnean name of *Gadus* has not been retained for any of these lesser divisions of the family.

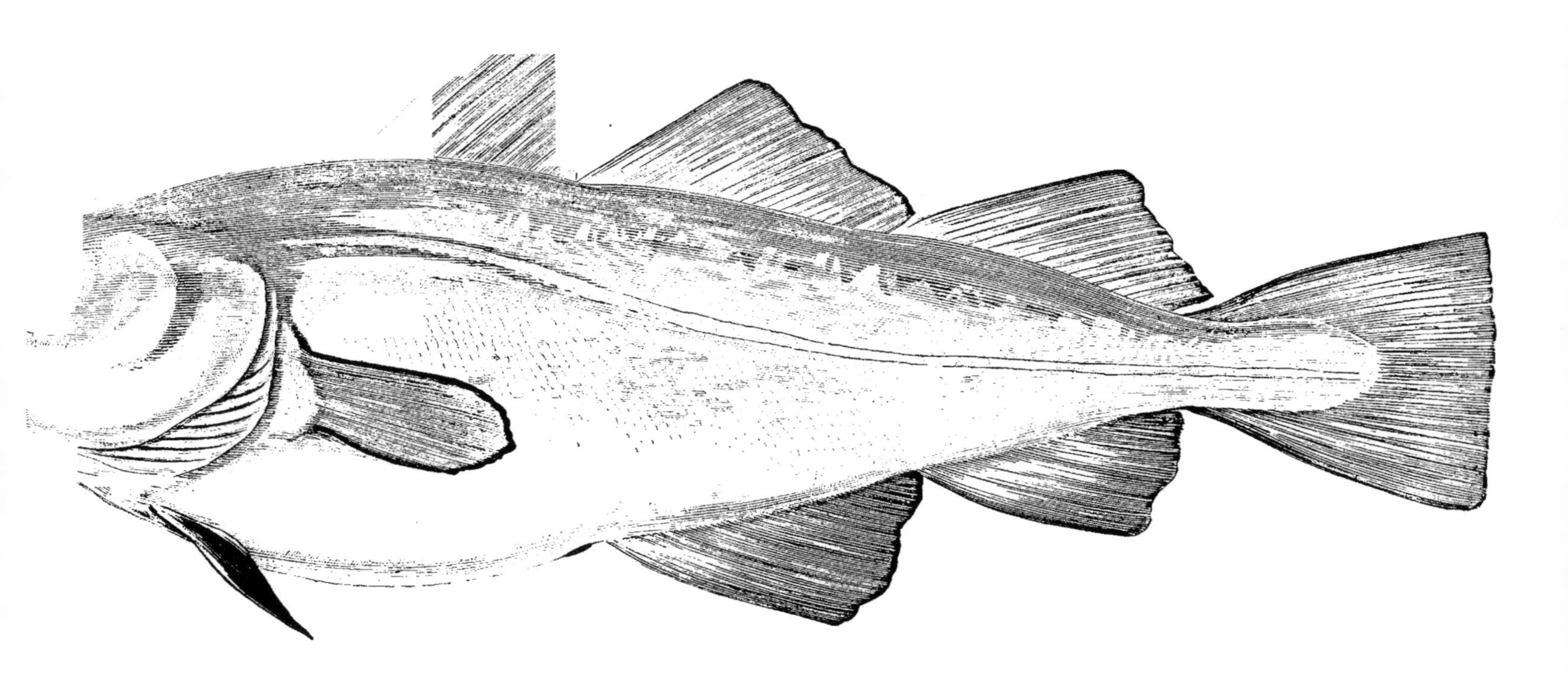

COD.

MORRHUA.

With the characters of the family as above, there are three dorsal and two anal fins; a barb at the point of the lower jaw.

COD.

Asellus cabelian,	Jonston; tab. 1, f. 1, and tab. 2, f. 1.
" "	Willoughby; p. 155, tab L. m. f. 4.
Gadus morrhua,	Linnæus. Bloch; pl. 64.
" "	Donovan; pl. 106.
Morrhua vulgaris,	Cuvier. Fleming; Br. Animals, p. 191.
" "	Yarrell; Br. Fishes, vol. ii, p. 221.
Cade morue,	Lacepede.
Gadus morrhua,	Jenyns; Manual, p. 440.
" "	Gunther; Cat. Br. M., vol. iv, p. 328.

The fishes of this family are of special importance in that they supply an abundance of wholesome and acceptable food to man; and so much the more valuable are they that they may be preserved with salt without being deprived of their good qualities; and in that condition they are conveyed to countries in which otherwise there might be a considerable deficiency. All the species are limited to the temperate and colder regions of the ocean, with a slight exception; and of them all the Cod is the most abundant, as it is also, in an economical point of view, the most valuable; so that in pursuit of it extensive fisheries are carried on in different parts of the world.

Within the range we have specified it is very generally scattered about, at the variations of depth from ten or twenty fathoms to fifty or sixty; but there are favourite districts in which they assemble in vast multitudes, and to which they are drawn chiefly by the abundance of acceptable food, but partly also by the nature of the ground, their chosen haunts being on the elevated surface of some subaqueous mountain or plain.

Their food is sought for on the ground, as is the case generally with fishes that are furnished with a barb below the chin; which organ is not a mere appendage, but by dissection is known to be supplied with a special nerve of considerable size, by which it is rendered so sensitive that on some occasions it appears to serve as a substitute for the absence of another chief organ of perception. I possess the note of a Cod of full growth which had swallowed a bait to be taken with a line, and which bore the appearance of being well fed, but which was altogether without eyes; and from the structure of the skin which covered the sockets there seemed no reason to suppose that it had ever enjoyed the benefit of possessing these organs. This appendage must therefore have well performed the office of a substitute, in conformity with the observations of Sir Charles Bell, in his Bridgewater Treatise on the Hand, (p. 48,) where he says, that by an anatomical investigation and experiment, he had discovered that the sensibility of all the head, and of its various appendages, is derived from one nerve only of the ten which are enumerated as arising from the brain, and are distributed within and around the head; and pursuing the subject by the aid of comparative anatomy, he found that a nerve corresponding to this, which is the fifth nerve in man, served a similar purpose in all the lower animals. In creatures which are covered with feathers or scales, or protected by shell, this nerve becomes almost the sole organ of sensibility. It is the development of this nerve that gives sensibility to the cirri which hang about the mouths of fishes. We may add that this fifth pair of nerves is represented in Monro's twenty-first plate of the Anatomy and Physiology of Fishes. The nerves of smelling also are well marked in these fishes, and derive their origin on each side from a round ganglion which is connected with the brain by a lengthened cord. The last-named writer also represents the organ of hearing as being distinctly recognised; so that the exercise of every sense is well provided for. We shall find more to remark on this subject when we speak of the genus *Motella*, an aberrant form of the Gadoid family.

The Cod, thus supplied with the organs of sensation, is one of the most voracious of fishes, and on most occasions appears to feed indiscriminately. Yet there is proof that it exercises decided preference for particular objects; so that it is not only caught with some baits in greater abundance than with others, but there are animals likely to be found in its stomach, while there are others which it is in vain to look for; which still are of common occurrence in other fishes of the same family which also gather their food from the ground. In addition to several sorts of bivalve shells, and one or two species of aphrodite, stones are found, of no small size, that have been swallowed because of the encrusting lepralia or corallines that covered them; and when the latter have been digested, the stones are probably rejected from the stomach. In one instance six Picked Dogfishes, each nine inches in length, were found in the stomach of a Cod; and the following list of crustacean animals (crab and lobster kind,) in the stomach of these fishes, which were taken in the west portion of the British Channel, will shew the strong preference which the Cod manifests for that sort of food; of which also, we may add, their digestion is so powerful and speedy, that, in a short time after being swallowed, the hard and brittle crust of the crabs is made so soft by the action of the gastric juice, that their legs may be twisted round the finger.

Crabs.—Stenorynchus phalangium, Achæus Cranchii, Inachus Dorsettensis, I. dorynchus, I. leptochirus, Hyas coarctatus, Eurynome aspera, Xantho tuberculata, Cancer pagurus Portunus corrugatus, P. arcuatus, P. marmoreus, P. pusillus, P. longipes, Gonoplax angulatus, Atelecyclus heterodon, Corystes cassivelaunus, Pagurus Bernhardus.

Long-tailed Crustaceans, Lobster kind.—Galathæa squamifera, G. strigosa, G. dispersa, G. Andrewsii, Munida Rondeletii, Gebia stellata, G. deltura, Nika edulis, N. Couchii, Squilla Desmarestii, Alpheus ruber, Scyllarus arctus.

In this enumeration the notes of Mr. W. Laughrin, A.L.S., are united with my own; and of these species the *Scyllarus arctus* offered only one example, which is now deposited in the British Museum; but of the *Munida Rondeletii*, which is usually considered as not a common species, there has been found not only numerous specimens, but these have often been of remarkable size. The longest leg of an example described by Mr. Bell in his beautiful Natural History of this tribe measured six inches, but I have found the same part to measure nine inches, with the antennæ of the same length as the leg. We find Cods which have been rioting on this crustacean food to be in good condition for the table; but I have been informed by an intelligent man who for several years had been employed in the fishery on the banks of Newfoundland, that such was not thought to be the case in that part of the world; but that, when the fish there were found to have their stomachs filled with crabs, although the fish were numerous, large, and appeared to be well fed, it was the usual practise for the ships to change their quarters in search of others.

The Cod sheds its roe in December and January, and as the grains are increasing in size the fish is in the best condition for the table; but its excellence has fallen back by the time the roe is ready to be shed, and after spawning this fish becomes emaciated and worthless. Indeed there is no fish, except perhaps the Salmon, that offers so great a contrast to itself from the time of its highest perfection to the worst, which is presently after spawning, and from which it is not speedily restored.

The Cod is one of the most prolific of fishes, as may be supposed when we call to mind the vast numbers which are caught at the principal fishing stations through a long succession of years, where one man in Newfoundland has caught five hundred and fifty-two in a day, and upwards of fifteen thousand in a voyage. Ten thousand Codfishes were reckoned a proper yearly capture for a man. The fact is well borne out by an examination of the multitude of grains of spawn which have been counted in the mass of the ovaries. In a fish which weighed twenty-one pounds, the roe weighed eleven pounds, or more than half of the whole bulk; but in another which weighed thirty pounds the roe weighed only four pounds and a quarter; and yet in this last instance the following proportion was fairly calculated. In repeated trials, two grains in weight of this roe gave the number of ova four hundred and twenty-three; so that, making a full allowance for the membrane mingled with them, the number of living individuals which might have been produced from this fish, in which the roe was of less than usual proportionate weight, was little less than seven millions. That very many of these eggs never reach a useful size is highly probable, and yet it is to be remarked that a young Cod is more rarely found in the stomach of other fishes than the generality of its fellow natives of the deep.

I have found the young ones less than an inch in length by the end of May, and they require at least the second year to render them fit for the market. When about half grown they are often caught in rough ground near the land, and being somewhat variously coloured they have been regarded as a separate species, with the name of Tamlin Cod, a name which is first mentioned by Jago, and after him by Borlase. The word *tam* in a western dialect signifies what is short and stout, and is applied to this fish because it is in full condition at the time when the full-grown fish has become thin from the effect of spawning.

Besides that the Cod in its season is an excellent dish for the table, an important use of it is when it is salted and dried, the ordering of which is effected in a different manner in different places, but into the peculiarities of which this is not the place to enter. The tongues and sounds (or air-bladders) are also preserved in pickle, and in this condition when boiled they form a very acceptable dish. The usefulness of this fish has also been more widely extended of late by the employment of the oil extracted from the liver in several diseases, in some of which it is found an important remedy. The first notice we have of it as such, was in the medical works of Dr. Bardsley, who mentions it as prescribed for chronic rheumatism in the Dispensary at Manchester, at the beginning of the present century, and since that time it has found general acceptance in glandular diseases among medical practitioners. But in England at least it has been found that this oil can only be extracted from the liver when the fish is in its best condition; for when its strength has become exhausted by the process of spawning, and until the recovery of its flesh, which is not soon effected, the liver is found in a like state of emaciation with the flesh, and affords no oil. We believe that the greater portion of what is now used is obtained from the fishery of Newfoundland, where, however, it has been said that the fish is scarcely equal to those caught in our own waters.

The value which has been set on the Cod is of comparatively modern date, since it does not appear that this fish was known at the tables of the Romans of the Empire; a circumstance to be explained by the fact that it was not found in the Mediterranean. The first regular fishery appears to have been carried on in the German Ocean, and that in very early times, as appears from the fact that it was so recognised before the year 1368, when the city of Amsterdam procured permission from the King of Sweden to form an establishment for carrying it on in the Isle of Schonen; and in the year **1415**, Henry the Fifth of England compelled the King of Denmark to make satisfaction to some of his subjects for injuries received in his dominions for something connected with it. Whatever was the nature of the privilege thus claimed, it was afterwards lost, until Elizabeth recovered it. This fishery was the principal source of the supply of Cods, until the discovery of the much larger numbers to be obtained on the banks of Newfoundland; when the attention of fishermen became directed to that more distant but more promising source of wealth, under the direction and with the assistance of merchants who made it a portion of the traffic which they were accustomed to carry on with the Italian ports of the Mediterranean. On these fertile banks the mode of fishing has varied, but it is only of late that reports have been circulated of a decrease in the numbers of the fishes that are found in that district; as if the long-continued and mighty inroads which have been made on them have at last effected a decided diminution of what may have appeared an inexhaustible supply. But Cods have long been found in large numbers nearer home, as on the Dogger Bank, and in our North-eastern Sea, where, along the borders of Northumberland and Norfolk, the fishery engages the service of a large number of boats and men, of which the port of Barking in particular affords an instance. According to evidence produced before a committee of the House of Commons, there belong to that place about one hundred and twenty fishing vessels, of the burden of forty to sixty tons, with a crew of upwards of eight hundred men; and their employment in this fishery lasts for about three months in the year, during which they are accustomed to make three voyages on the whole.

But within a year or two a new discovery has been made of a situation, which for a time, is likely to draw to itself the attention of fishermen of the northern portion of our island and perhaps of Ireland, in a higher degree than any other. This is along the upper portion of a submarine elevation, of which the situation is marked by a solitary rock that bears the name of Rockall, and which probably was better known at some distant date as the resort of fish, than more lately, down to the time of its renewed discovery. In the meanwhile the fish has had time to grow, as well in size as numbers, so that wonders are told of the success of the first adventurers to the spot; from which two boats returned after a week's fishing, each with between thirteen and fourteen tons of fish. The size of the individual Cods is not mentioned, but as a single example was known to have lived in an enclosed pond at Logan, in Scotland, to the supposed age of about fifteen years, during which it is said to have made a gradual increase in bulk, we may judge that those taken at Rockall, at freedom and fully fed, had attained to the full of that which at any time they reach. A successful fisherman on the banks of Newfoundland informed me that out of many hundreds he once caught there, there was a Cod which reached to a hundredweight, and that with a wish to show it to his friends at home, he purchased it of his captain for the price of half-a-crown. The largest Cod I have known weighed fifty-six pounds; but scarcely any are in finer condition than those which abound in the deeper water between the Scilly Islands and the west coast of Cornwall, and also between St. Ives on the north and the Mount's Bay.

The fishery for Cods is conducted with hooks, and either with a single line from the boat, (each fisherman attending to a couple,) or with long lines, which in the west of England are termed bulteys, or bulters, and which cannot be shot in such deep water as may admit the single line. These bulteys are formed of a principal line, which is a stout cord or small rope, and to which is fastened a series of short lines about a fathom in length, placed at such distances from one another as that they shall not be entangled together. Sometimes many hundreds of these hooks are thus fastened together, with a stone or grapnel to moor them, and with a cork-line to mark the place and draw them up. The baits are various,—as Herrings, Pilchards, and Lamperns; and the direction is across the course of the tide, on ground where the hooks are not likely to get entangled amidst the rocks. The whole is drawn up at such a time as experience has taught the fishermen to be sufficient for their purpose. If left long after the fish are dead they are subject to the depredations of some of the sessile-eyed crustacean animals, termed by fishermen sea lice;

which enter their bodies by the mouth and gills, and in a time surprisingly short devour the whole of the soft parts, so as to leave the skin almost empty. Of this last-named method of fishing the success must be greater than any which can arise from the employment of a few lines that hang from a boat which is manned by no more than two or three up to half a dozen men; but it requires a greater outlay than many fishermen are able to provide, and a complaint also is sometimes made of the want of bait for such a multitude of hooks. But several hundreds of fishes, including the Cod and Ling, are thus sometimes drawn up at a single haul, and that too at times when boats which must ride at anchor with their lines are not able to encounter the roughness of the sea.

It has been observed that the largest number of these fishes are often caught when the sea is becoming rough with the threatening of a gale from the direction of the deeper sea, yet a heavier storm is said to drive them away. When not sold fresh these fish are prepared with salt for exportation, and also for consumption at home; for which purpose the head and a portion of the backbone, with the entrails, are removed, in which condition they are salted and dried. In the year 1853, according to a report of the Board of Fisheries, the quantity of Cods, Ling, and Hakes cured in Scotland and the Isle of Man, amounted to somewhat more than five thousand nine hundred tons; to which are to be added upwards of three thousand tons which were sold fresh, the whole amounting to nine thousand three hundred and forty-two tons and five hundredweight; but this was the highest that had ever been known. Large quantities of Cods which have been thus prepared in Newfoundland are consumed in England. On a copper coin struck in or for the Magdalen Islands, in the Gulf of St. Lawrence, with a seal on the obverse, the reverse bears a Cod split and prepared in the manner we have described.

The Cod is the stoutest species of this family in proportion to its length. The head large, but in a fish in good condition the outline rises from the snout to the beginning of the dorsal fins. The upper jaw projects a little beyond the lower; teeth in both, and a plat in the form of a horse-shoe in front of the palate; a barb on the under jaw. Eye moderate. Body slightly compressed at first, more so behind the vent to the tail; vent midway between the snout and root of the caudal fin. Scales small; lateral line conspicuous, at first nearer the back, lower and straight behind. Dorsal fins three, the first beginning a little behind the root of the pectorals, irregularly rounded; anal fins two; pectorals round; ventrals short, the first rays extended and pointed; tail slightly round. Colour on the back dark yellow, sometimes brown; sides mottled; belly white; all the fins soft. The Cod of the north of England has the snout much shorter and rounder than the fish of the west coast. We have already mentioned the great weight to which it sometimes reaches, but from thirty to forty pounds is a more general size.

From Griffith's translation of Cuvier's "Animal Kingdom" we learn that in the neighbourhood of the Isle of Man, there is found a variety of the Common Cod, which is named Red Cod, or Roek Cod, the skin of which is a brightish vermilion colour, and the flesh of it is considered superior to that of the other. It was the opinion of Dr. Turton that there was also another species of British Cod, which he named the Speckled Cod, (*Morrhua punctata*,) but which he himself appears not to have met with when he published his translation of Linnæus's "System." It has been shewn, however, in a way not to be doubted, by Dr. Dyce, in the "Annals, etc. of Natural History," 1860, that this supposed species was no other than a mis-shaped example, such as is scarcely uncommon among fish, of the Common Cod; and Dr. Dyce illustrates his observations by some dissections which prove that the foundation of the deformity was in the structure of the bones of the back, as was the case also in the deformed example of the Codfish which we have described.

A Cod came under observation, which was in good condition and of full growth, which possessed only one pectoral fin, while on the other side there existed merely a stump, which had the appearance ot having been originally formed in that condition. It is probable that it was in consequence of this deficiency the ventral fins had been called more particularly into action, for the purpose of regulating the positions of the body.

The *Lernæa branchialis*, which is a large and formidable parasite, is not unfrequently found firmly attached to the gills of this fish.

HADDOCK.

Callarias,	JONSTON; Table 1, f. 2.
Asellus antiquorum,	WILLOUGHBY: Table L on the plate, but *Onos sive asinus antiquorum,* and *æglefinus,* the Haddock, p. 170.
Gadus æglefinus,	LINNÆUS. BLOCH; pl. 62.
" "	DONOVAN; pl. 59.
" "	JENYNS; Manual, p. 441.
Gade æglefin,	LACEPEDE.
Morrhua æglefinus.	CUVIER. FLEMING; Br. Animals, p. 191.
" "	YARRELL; Br. Fishes. vol. ii, p. 233.
Gadus æglefinus,	GUNTHER; Cat. Br. M., vol. iv, p. 332.

WITH an approach to the same form and organization with the Cod, the Haddock comes near to it also in its habits; but although equally dispersed over the world, it is for the most part in less numbers. It is observable also that this fish is disposed to observe a partial and limited migration or change of quarters, with a somewhat loose arrangement of the multitudes that observe it. Such is the case in a remarkable degree on the coast of Scotland; and also on a portion of the coast of Yorkshire, where there is a bank which extends for about eighty miles, but in breadth scarcely exceeding three, where in the winter they are caught in large numbers; but on either side of these limits at the same time none are taken. It is also found in abundance in America, on the borders of Massachusetts; but it so little affects the society of the Cod, that on the banks of Newfoundland, when a fisherman succeeded in taking upwards of five hundred and fifty Cods in one day, he took no more than two Haddocks at the same time.

In their periodical assemblings at their favourite stations on our coasts they appear to be influenced by a common feeling, which may be of the same nature as that which prompts

HADDOCK.

them finally to the development and shedding of the spawn, the season of which, as is generally the case with the fishes of this family, is in the colder months of the year; and after continuing in numbers for about two months, during which they have yielded to the fisherman an abundant harvest, they go away into deeper water or a colder zone; and although single examples may be caught at any time, the greater number does not show itself again until the return of another season.

The Haddock is in sufficient estimation for the table as to meet with a ready sale; but neither in numbers nor as food is it equal to the Cod, whether fresh or salted; and as regards the last particular, there is much difference of opinion whether it deserves the credit in which it stands; but this difference may in a great degree depend on the nature of the district in which the fish was caught, as well as in the sort of preparation to which it has been subjected. It is admitted, however, that the older and larger examples are inferior to such as are of moderate size. It is chiefly in Scotland that the salted Haddock is of sufficient importance to be the subject of trade, and a few of the towns in that portion of the United Kingdom have obtained some degree of celebrity from the manner in which these preserved fish have been prepared. Such is the case with Findhorn, which has secured a reputation on this account, which is more than shared by some other places in its neighbourhood, although less generally known. The principal portion of the secret in the preparation of this esteemed dish is said to consist in smoking the fish over a peat fire after it has been for a short time moderately salted.

The Haddock feeds near or from the ground, and uses little discrimination in the choice· and yet, while it rejects nothing which the Cod might swallow there seems to be that difference of appetite between these fishes, that the stomach of the Haddock will best repay the examination of the naturalist whose interest is in the collection of shells, of which he will thus secure some species that otherwise he might not readily meet with. In a single stomach, among a multitude of univalve and bivalve shells, I was able to select no less than twelve separate species.

There are at times some unknown influences in the ocean

which have caused great destruction among the multitudes of these fish, so that large numbers have been found dead and floating on the surface of the water. An instance of this sort is recorded in the parliamentary inquiry into the state of the Salmon fisheries in the year 1825; from which it appears that about thirty years before that date so great was the havoc among these fishes, that ships had sailed through many leagues of the North Sea where the surface was covered with dead Haddocks, and after this for several years it was a rare fish in these districts. It was seen also that even when again they had become plentiful it was long before these fishes had reached to their former size.

It is not often that the Haddock attains the length of two feet, or exceeds the weight of eight or ten pounds, but Mr. Thompson mentions instances of examples taken in Ireland which were of eighteen, twenty, and twenty-five pounds; and when this gentleman adds that these fishes in Ireland are often valued more highly than the Cod, and obtain a price which *we* should deem enormous, we are driven to the conclusion that these Irish Haddocks are more richly fed, and in finer condition than in most parts of England. The higher price cannot proceed from a scarcity of the fish, for it is said that in Dublin Bay and along the neighbouring coast they are in great plenty.

The head is compressed, level on the top, with a ridge, which is directed backward. Snout projecting, nostrils half way to the eye, which is large, elevated, and behind the corner of the mouth; the jaws are nearly equal, but the upper jaw is within the projecting snout; teeth in both, and in the palate: a barb at the lower jaw. The body compressed, rising from the head to the first dorsal fin, more slender towards the tail. Vent about midway between the snout and root of the tail. Lateral line nearly straight, conspicuous from its dark colour: scales on the body slightly visible. Dorsal fins three, the first elevated, triangular, ending in a point; second and third less elevated, extending to near the tail. Anal fins two, the first forming the segment of a circle. Pectorals slightly pointed; ventrals with the first ray lengthened; tail more or less concave. The colour of the back and fins dusky purplish brown, paler on the sides, dull yellow or white below; a large dark spot on the side, which is lighter in the middle, stretches down from the lateral line. The Haddock of which a figure is given (Pl. 19) in Fries and Eckstrom's "Skandinavian Fishes," is so unlike the British species as to raise the suspicion that they may be specifically different; and the same may be said of the Pollack.

Naturalists of former days were persuaded that the names Onos, in Greek, and Asinus, (the ass,) were the proper designations of this fish in ancient times; and when we examine the colour it usually bears, coupled with the distinguishing stripe at the shoulders, we scarcely feel surprised that the excellent naturalist, Turner, in the age of Queen Elizabeth, and the still more eminent Ray, should countenance this opinion. But it happens unfortunately for this idea that the Haddock is not found in the Mediterranean, and therefore could not have fallen under the observation of the Greeks, from whom the appellation was borrowed by the Romans, as applied to a species with which both these people were acquainted. To a kindred fish therefore this name must have been first applied; and in the Hake we shall find sufficient likeness of colour to the terrestrial animal, to warrant the comparison by a people of whom we are constantly reminded that with them a distant resemblance was sufficient to constitute a likeness that would authorize a name; but whether this Hake of the Mediterranean is the same with that known among ourselves remains yet to be determined.

DORSE.

VARIABLE COD. BALTIC COD.

Asellus varius,	WILLOUGHBY: p. 172, table L. 1.
" "	BLOCH; pl. 63.
Gadus callarias,	LINNÆUS.
Gade callarias,	LACEPEDE.
Morrhua callarias,	CUVIER. YARRELL; Br. F., vol. ii, p. 231.

THE Dorse is especially a fish of the north, for it exists in large numbers in the sea about Greenland, and even in the Frozen Ocean on the north of America. It is familiarly known also on the coast of Norway and Sweden, and further in the Baltic, where it is said to ascend rivers as far as the tide reaches; but its appearance further south is uncommon, and it is only as a rare straggler that it has shewn itself in the west of England or south of Ireland. Yet I have known a few instances in which it has been taken on the north and south borders of Cornwall; and I feel little doubt that what has been supposed a rare variety of the Haddock, preserved in the museum of the Dublin University, and referred to in the catalogue of that collection, as also mentioned in Mr. Thompson's "Natural History of Ireland," is in fact an example of the Dorse. The mistake here supposed, of confounding this fish with the Haddock has occurred in at least one other instance, and is the rather to be excused as in its general form it bears a nearer likeness to the last-named fish than to a well-fed ordinary Cod, although indeed in colour it differs greatly from both.

The peculiar habits of the Dorse as distinguished from the others of its family have not been communicated to us; but we know it takes a bait, and as food it is said to be of superior quality, this preference being assigned to it even after it has been salted.

DORSE

CXXXVII

The communication which follows is from a gentleman who has made himself known to naturalists by the name of Piscator.

"Launceston, September 7th., 1843.

"Dear Sir,—Whilst observing a man who was casting his line from the rocks near Boscastle yesterday, I perceived him draw on shore a small species of Cod, about six or seven inches long, different from any of the Cod tribe I had before seen, but exactly answering the description and the plate of the Dorse, or Variable Cod, in the second volume of Mr. Yarrell's "British Fishes," except that instead of being spotted, as there described, it was of a dark coppery tinge we so often see in young Whiting Pollacks when caught on rocky ground. At a first glance indeed I thought it was a small Pollack, till the short lower jaw and beard below instantly pointed out a distinct species. The eyes were large, irides of a golden hue, the pupils intensely black and sharp, and remarkably brilliant; the nose is prominent, projecting a little beyond the upper jaw, like that of the Haddock; the under jaw short; the upper part above the lateral line of a deep chocolate cast, assuming a coppery tinge along the sides, and becoming much paler towards the belly; the lateral line very distinctly marked and silvery, rising in a curve over the pectoral fin, then descending and passing in a straight line to the tail. Caudal fin square at the end, of a darker colour, assuming a dusky tinge, as indeed were all the fins. I have been thus particular in my description, which so far corresponds with that of Mr. Yarrell, that I have no doubt but that it is the same identical Dorse as described by him. The colour, we all know, of fishes inhabiting rocky ground can never be relied much upon, and varies exceedingly in different specimens taken even in the same spot. I considered this communication might not be uninteresting to you, more particularly as Mr. Yarrell mentions that the authority upon which the Dorse was originally introduced seems now to be questionable, though it is well known in the Baltic, and frequently called the Baltic Cod. But from the juvenile appearance of the specimen I had an opportunity of seeing, there can be little doubt that it was a native of our own coasts. Mr. Yarrell mentions he had never seen a specimen. I wish I had thought of preserving mine in spirits: now, from the heat of the weather it is spoiled by decomposition."

Since the date of this letter I have been informed of an example which was obtained in St. Austle Bay, on the south-east coast of Cornwall; and two specimens have been caught by fishermen of Polperro, from which our figure and description, with additional notes, have been taken. Added to these, Mr. Thompson, of Weymouth, informs me that in the months of October and November, 1855, four examples were caught at one time, and ten at another, in trawl-vessels belonging to that port. They were of a golden yellow colour, and of small size, not exceeding three or four inches in length. Of this size indeed colour may not afford a decided mark of distinction, but the form of this fish is so different from that of the Common Cod, that no mistake needs to be committed in confounding one with the other.

The example selected for description was twenty inches in length. Compared with the Cod the snout projects considerably more, pointed, bent down, cavernous; flatter than the Cod backward from the snout and between the eyes; under jaw much shorter; the eye large, brilliant, even with the top of the head. Behind the head on the back a deep chink, almost like that on the nape of the Rockling, but without a ciliated membrane, as in that fish. Body like that of a Cod (or Haddock) lateral line conspicuous bent down at half its length. Most of the fins more expanded than in the Cod; the third dorsal and second anal running near the tail, and liable to fold down; tail round; the fin rays stouter in proportion than in the Cod. The colour much varied, the ground-colour yellowish or orange, with mottlings; back rich light brown; fins reddish yellow, mottled; some green tints on the sides; belly pale white; but all the colours disposed to fade. Barb at the lower jaw prominent. Fin rays,—of the first dorsal fourteen, second and third dorsal each with seventeen rays, pectoral eighteen, ventral six, first anal nineteen, second anal seventeen, caudal thirty-four. In its stomach was a crab, (*Zantho florida.*) Schonfelt is quoted as saying that when kept in a pond the Dorse devoured the smaller fishes.

The other example was taken in the company of Haddocks in March, as the former had been in December; its colour a bright golden yellow, paler on the sides and belly. Lateral line at first of the colour of the skin, but more golden nearer the tail. The connecting membrane of the fins, uniting the rays, diaphanous from the roots, so that the rays themselves could be easily traced.

Mr. Thompson mentions what was supposed to be a Haddock, obtained by Dr. Ball, at Dublin, of a canary colour; and another where the upper surface—the head and back—were of a pale golden yellow, which extended to the dorsal, caudal, and pectoral fins, without the mark on the sides common in the Haddock. Other examples, much like these, but more varied are mentioned, and one of them twenty-seven inches in length; and scarcely a doubt can remain that these fishes were examples of the Dorse. Willoughby mentions one caught by a fisherman which measured four feet, which, he justly observes, was an uncommon circumstance, and the rather so as it has hitherto remained the only authentic instance of the capture of this species in Britain.

Dr. Gunther represents this as only a variety, or the young condition of the Common Cod; but I have seen an example of the Dorse, as described above, as large as an ordinary Cod, and easily to be distinguished from it; as also I have examined Codfishes, even of minute size, (down to the fourth of an inch in length,) but the general shape of which was decidedly different from the fish I have here represented.

BIB.

WHITING POUT. BLENS. BLINDS.

Asellus luscus,	WILLOUGHBY; p. 169, table L.
Gadus luscus,	LINNÆUS. BLOCH; pl. 166.
" "	DONOVAN; pl 19
Gade tacaud,	LACEPEDE.
Morrhua lusca,	CUVIER.
" "	FLEMING; Br. Animals, p. 191.
" "	YARRELL; Br. F., vol. ii. p. 237.
Gadus luscus,	JENYNS; Manual, p. 442.
" "	GUNTHER; Cat. British Museum, vol. iv, p. 335.

THE Bib is known along the whole of the coasts of the United Kingdom, but is scarcely common in the north of Scotland and Ireland. On the other hand, it is found in the south and west of the last named country and England through the year, and at times, especially in the autumn and winter, it is even abundant. Its chief resort is in rocky places, where it finds its congenial food in the multitude of crustaceous animals and small fishes which frequent such neighbourhoods, but sometimes they pass into gullies and recesses where the bottom is irregular or formed into pits. In general the food is sought for at an higher elevation than is usual with the Cod and Haddock, and consequently what is found in the stomach is of a different kind.

The spawn is shed towards the end of winter, and, perhaps, generally later than in several others of this family of fishes. Considerable numbers are sometimes caught with a line, but although good as food, they do not stand on equality with the Cod or Whiting, and they are supposed to suffer decomposition more speedily than these fishes. When drawn up with a line it is common to find the transparent covering of the eye inflated

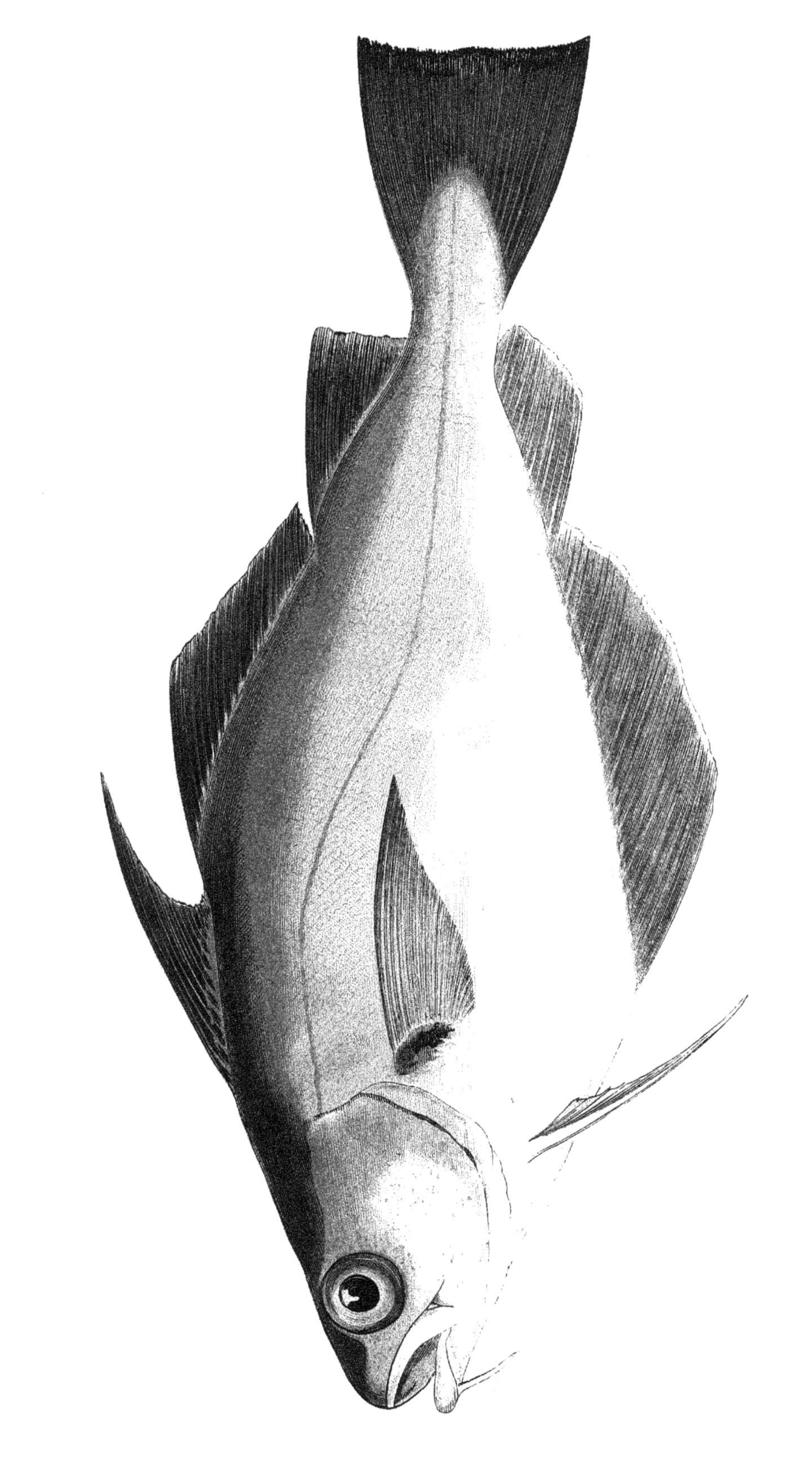

into a bladder, and even the dorsal fins are often distended in the same manner. It appears to be caused from the terror of the fish, by the agony of which the air of the swimming bladder is driven into these membranous parts; and it is this circumstance, which in a less degree may be observed in other fishes, that appears to have given occasion to some of the names by which the Bib has been designated.

In shape this fish is the deepest of the British species of its family in proportion to its length, which does not often exceed a foot, although it sometimes exceeds this measure by a few inches. I have known it to weigh four pounds. The head and body are compressed, snout short and blunt, gape moderate, under jaw slightly the shortest, teeth in both, and in the palate; barb at the lower jaw. Eye rather large, not far from the snout: nostrils in a depression before them. The outline rises at first in a rounded form from the snout to the origin of the first dorsal fin, and does not begin to descend until it has reached the second dorsal, from which it slopes gradually to the tail. The greatest depth is at the vent, which is nearer the front than a third of the whole length, and almost under the root of the pectoral fin. Scales small, and easily lost: lateral line high at first, sloping down opposite the end of the pectoral fin. The first dorsal fin rises to a point, long enough to overlap a portion of the second. Pectorals pointed; tail slightly concave; first ray of the ventrals long and slender, reaching beyond the vent. Colour of the back reddish brown or dusky yellow; sides coppery, and so also in some instances the belly, sometimes also with irregular dusky shades. Not unfrequently the sides are marked with bands of a deeper colour. A black spot at the origin of the pectoral fin. A border sometimes light coloured, sometimes dark, round the extremity of the tail.

POWER.

Power,	JAGO; in Ray's Synopsis Piscium, p. 163, and a figure.
Gadus minutus,	LINNÆUS. JENYNS; Manual, p. 444.
Morrhua minuta,	FLEMING; Br. Animals, p. 191.
" "	YARRELL; Br. Fishes, vol. ii, p. 241.
" "	GUNTHER; Cat. Br. Museum, vol. iv, p. 335.

I AM not able to refer to Willoughby and Bloch for this fish, since their figures of ***Asellus minor*** bear little resemblance to the Power; and the fish called Capelan by Rondeletius, and by subsequent writers who refer to him, is still less like the British species.

The name by which this fish is known by us is believed to be an ancient form of the word poor; in proof of which we find in a document of the reign of Henry the Eighth, the words powre and pore used instead of poor; but this fish is inferior to others of this family only as being much smaller, on which account it is chiefly employed as bait for other fish.

There is much similarity in the habits of the Bib and Power, as there is also in shape. Both prefer rough or rocky ground, from which they do not wander to a considerable distance; and the last named fish maintains so close an attachment to its favourite haunts that fishermen have informed me they only leave the place, and that in a body, about the month of April, when it is supposed they have sought out a proper spot for the purpose of spawning. It is further observed that when the Bib and Power inhabit the same ledge of rocks they do not associate together, but the Powers keep at their stations lower down, where the rocks rise from the level, while Bibs prefer the higher portion. And again, while Bibs are sometimes found in submarine gullies, the Power is very rarely caught in the same situation. During the fishery for crabs they are not unfrequently found in the crab-pots; and in autumn they are

POWER.

CXXXIX

occasionally taken with a line from a pier or rock in harbours. Their food is the smaller crustacean animals, and they swallow a bait freely.

The common length of the Power is from six to eight inches. Head and body compressed, deep, but the latter less so in proportion than the Bib, the deepest part at the vent, opposite the termination of the first dorsal fin. In front of the eye the head is short, the profile rounded from the upper jaw. Eye large, a row of pores between the eye and upper maxillary bone; jaws about equal, with teeth in both, and in the palate; a barb at the lower jaw. Body clothed with fine scales, which hang so loosely that the fish can scarcely be touched without removing them. The lateral line begins high, and bends down beyond the end of the pectoral fin, from thence straight. Vent nearer the tail than in the Bib, opposite the termination of the first dorsal. The first dorsal also begins further distant from the head; the pectoral shorter; ventrals also shorter, not reaching half way to the vent; tail slightly incurved. Colour of the upper parts dusky or yellowish brown, sides lighter, belly white.

MERLANGUS.

THIS genus differs from *Morrhua* only in that there is no barb below the point of the lower jaw; but while such is the principal mark of distinction between them, it is to be remarked further, that in the genus *Merlangus* the body is more slender and better fitted for active motion; which is the more requisite as seeking their prey in a higher region, they have to pursue creatures which are endued with greater facility of escape than is the case with such as fall to the lot of those of their family which obtain their food at the bottom.

WHITING.

Asellus mollis,	JONSTON; pl. 2, f. 3.
" " *major, sive albus,*	WILLOUGHBY; pl. L. M. 1.
Gade merlan,	LACEPEDE.
Gadus merlangus,	LINNÆUS. DONOVAN.
Merlangus vulgaris,	FLEMING; Br. Animals, p. 195.
" "	JENYNS; Manual, p. 445.
" "	YARRELL; Br. Fishes, vol. ii, p. 244.
Gadus merlangus,	GUNTHER; Cat. British Museum, vol. iv, p. 334.

THE Whiting is common and sometimes abundant in the west of England and south of Ireland, and in these districts it reaches a size and perfection which are scarcely seen in the east or northern portions of the kingdom; while in the far north of Scotland it is rarely met with, as, on the testimony of Dr. Barry, is also the case in the Orkney Islands. This limited extent of wandering seems to imply that Whitings are more sensible to the feeling of cold than several others of this family; in further proof of which it was noticed that in the month of February, when they were on the coast in large numbers, and those of largest size were with enlarged roes and milts, on a change of wind to the north, accompanied with a fall of the thermometer from 47° to 44°, although the weather continued moderate, the whole of the larger of these fishes went

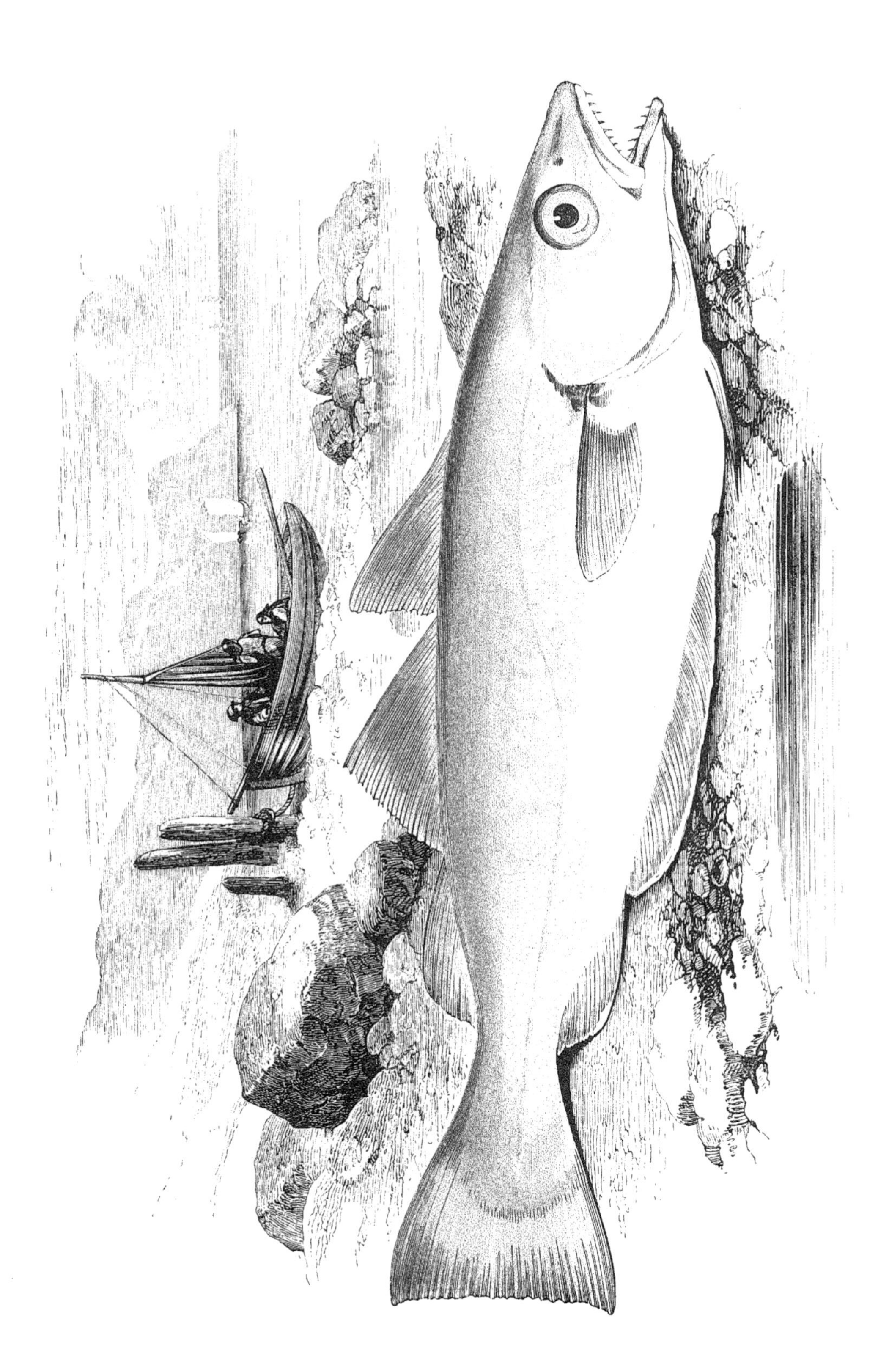

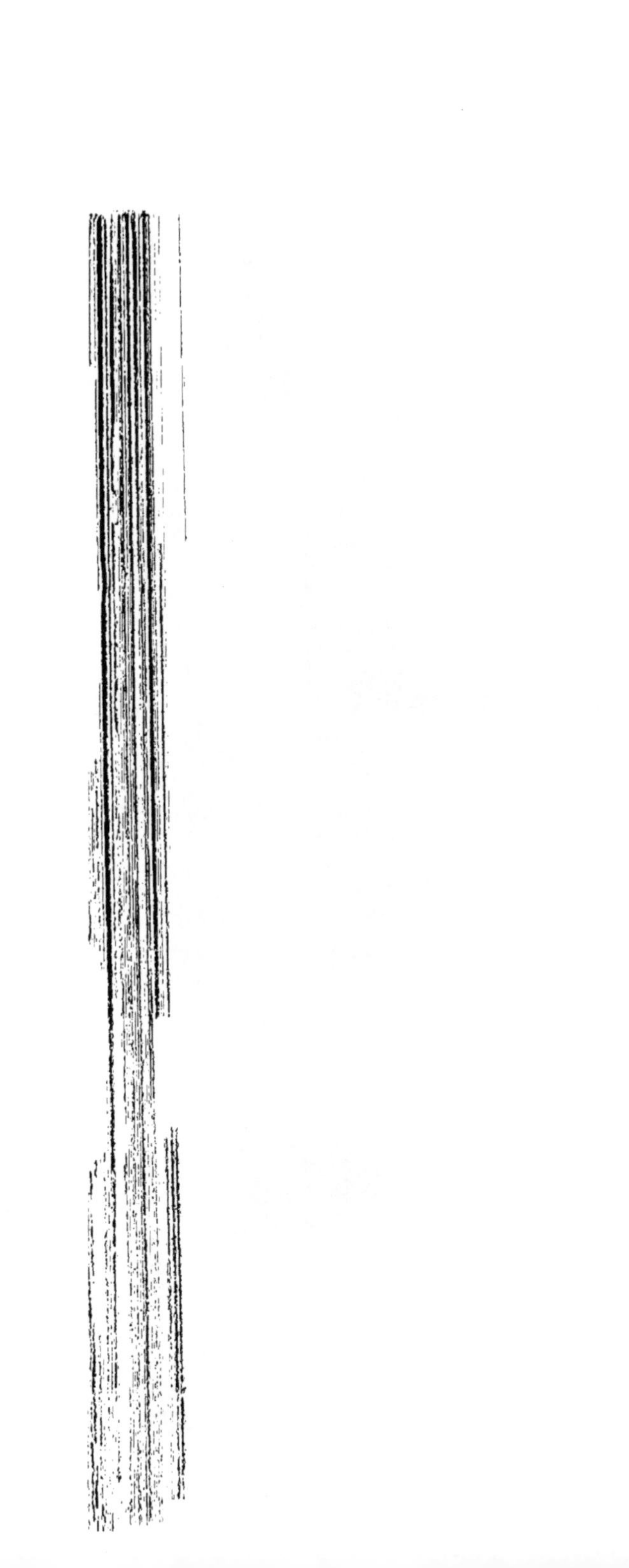

immediately into deep water beyond the reach of the fishermen, and only some very small ones remained. The fact that several sorts of fishes, although swimming at the depth of fifty fathoms, are speedily sensible of changes in the wind and weather, and even shew that they anticipate those changes by their motions and appetite, is well known to fishermen, and is proved in innumerable instances, even when such change is the opposite of what is here mentioned.

But there are other causes which incite this fish to change its quarters, and it appears to be rather in pursuit of prey than from sociability of disposition that they sometimes collect together in great numbers, which they usually do in sandy bays, or at least on level ground at no great distance from land; and when the young of several sorts of fishes abound, as they do especially towards the end of summer and in autumn, Whitings are busily eager in following them in all their movements. It is then also that they especially become the prey of the fisherman, whose most successful time is early in the morning and in the evening. The common mussel, or a slice of the cuttle (*Sepia,*) are temptations which cannot be resisted, but a preference would be given to a living prey, which is seized with indiscriminating voracity. From the stomach of a Whiting that weighed four pounds were taken four full-grown Pilchards.

The Whiting is in its best condition when the action begins which enlarges the roe, and which takes place between November and the early months of spring; but it suffers less than most others of the Gadoid family from the exhaustion caused by this process of nature, and hence it is in a condition for the table at all times. Few sea fishes are in higher esteem as food, and especially where the stomach requires what is easy of digestion; but it quickly suffers change, and therefore when delay occurs in the sale they are salted and dried, in which condition, when skilfully done, they are much valued. Willoughby says that in some parts of the continent an infusion of tumeric was employed to stain (yellow) these dried fish, in order to give them a richer appearance, and with the pretence also to make them more palatable. The smaller examples, when salted and dried, have received the name of buckorn. The abundance of Whitings has within a few years become considerably less than formerly from causes of which we may speak when we have to describe the habits of some other kinds of fishes, and especially the *Pleuronectidæ* or flatfishes.

This fish will reach to about sixteen inches in length, and a usual weight may be three or four pounds, but I have been informed of one that weighed seven pounds. Head and body compressed, the deepest part at the vent, which is opposite the middle of the first dorsal fin; eye moderately large; upper jaw a little beyond the lower; long sharp teeth in both, and a triangular spot of teeth on the palate. The scales small; lateral line high at first, and becoming lower opposite the end of the pectoral fin. Dorsal fins three, and two anal; pectoral rounded at its termination; ventrals slender; tail even. Colour of the back dusky yellow, the sides paler, but often with dashes of the same; belly silver white; edge of the anal fins whitish; a dusky or black spot on the upper side of the root of the pectoral fin.

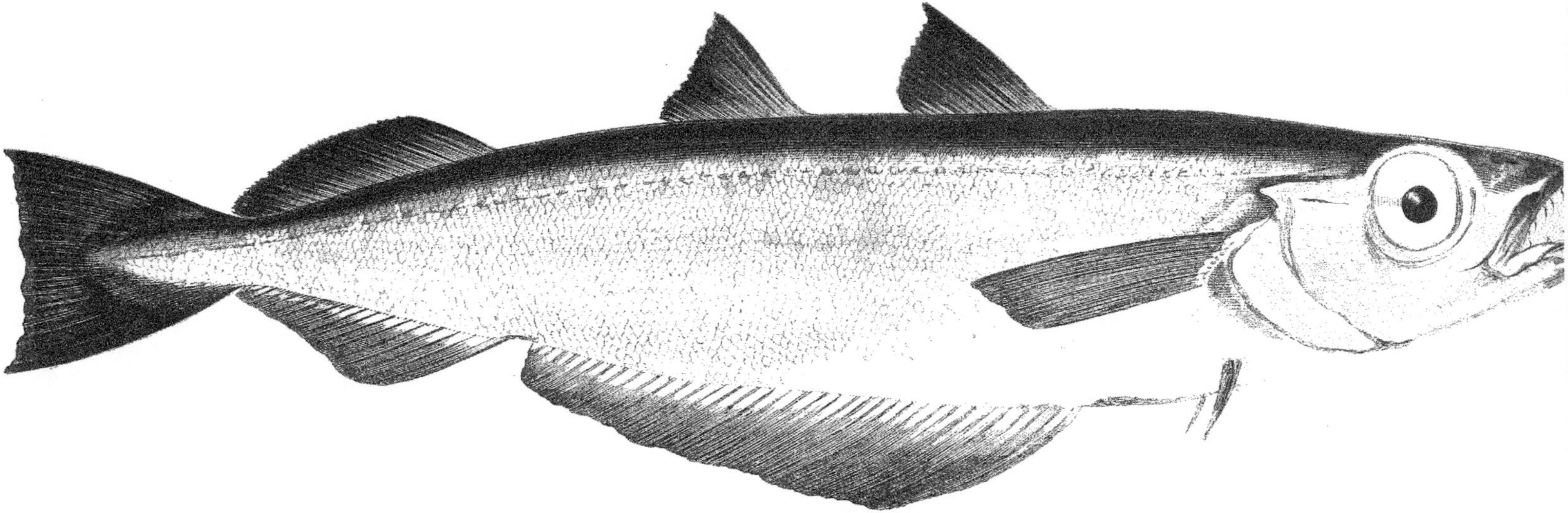

POUTASSON.

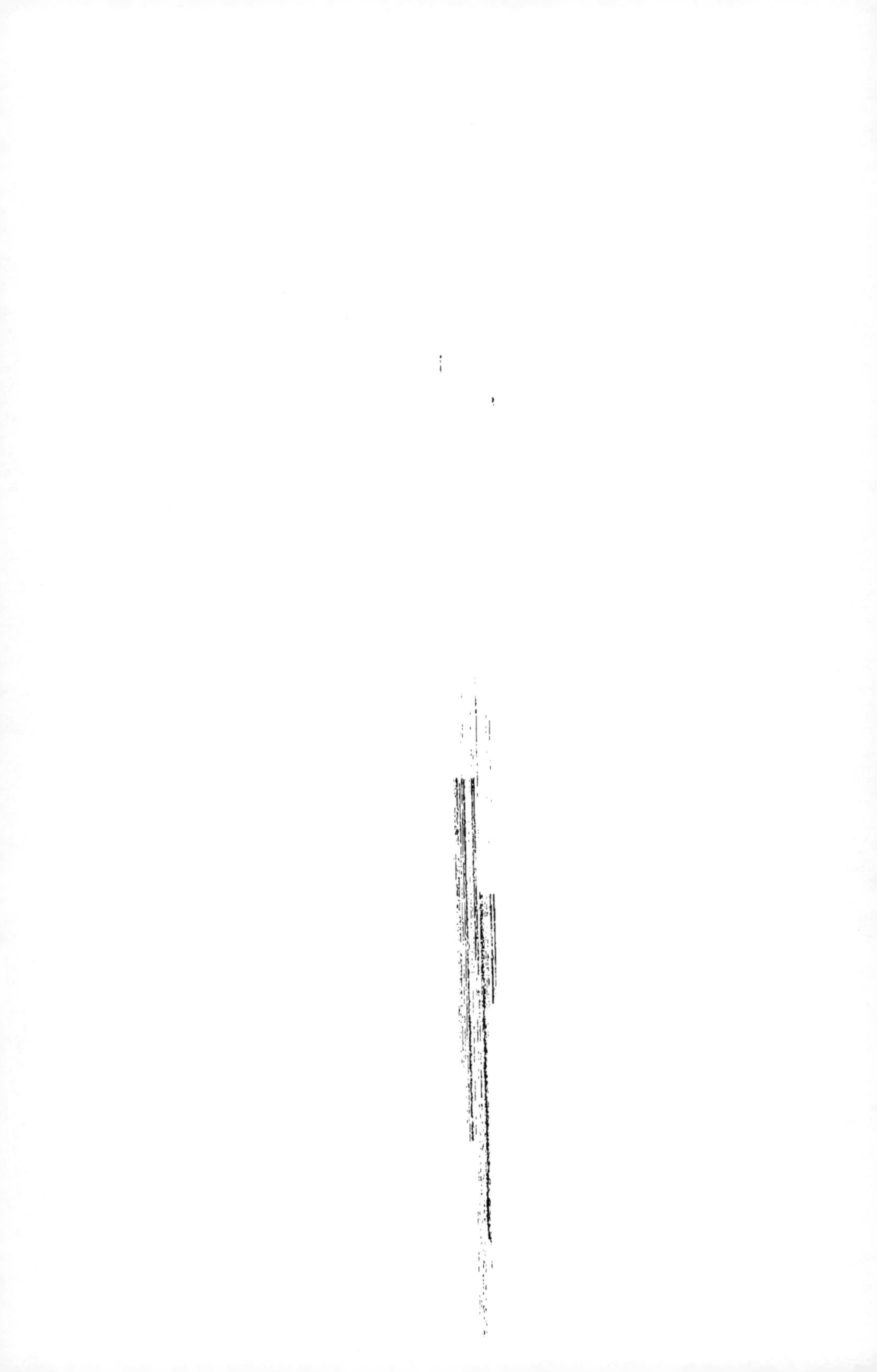

POUTASSOU.

Poutassou gros,	Risso; the Poutassou simply being the Pollack.
Couch's Whiting,	Yarrell; Br. Fishes, vol. ii, p. 247.
Gadus poutassou,	Gunther; Cat. Br. Museum, vol. iv, p. 333.

This fish was formerly mistaken for the Common Whiting, but Risso was led to suppose that there was a specific difference between them, although no opportunity had offered itself to him for comparison. In the month of May, 1840, an example answering to Risso's description of this species, was taken with a line by a fisherman of Polperro, and, as even to this ordinary observer, its difference from the well-known Whiting was apparent, it was brought to me for further inquiry. A description and figure were taken, and the first announcement of its discovery in Britain was made by Mr. Yarrell, in the second edition of his "History of British Fishes," where, however, the engraving represents it much darker than it should be.

Risso says that its haunts are in the deepest portion of the sea in the neighbourhood of Nice, where it is fished for at all times of the year; that it spawns in the spring, but that its flesh is rather soft. And this seems to have remained the whole which was known of its history, until the summer of 1851, when, in the month of July, I received information that immense numbers of small fishes were at a few miles from land along our coast, and that the larger fishes were devouring them eagerly, so that the stomachs of the latter were found distended with them. There was little difficulty in procuring an opportunity for examination, and I was agreeably surprised to find that these numerous small fishes were the young of the

Poutassou, which therefore must have produced them on our coast, although the parent fish in this instance had not been discovered. These young ones measured about five inches in length, and closely resembled the larger example, except perhaps that the body was more slender, and in consequence the head appeared proportionally a little larger. They continued to abound for about three weeks, at the end of which, between the 21st. and 23rd. of July, they all suddenly disappeared.

The length of the example first referred to was fifteen inches; depth two inches and a half, the greatest depth being at the vent, which is anterior to a line drawn from the origin of the first dorsal fin; from the mouth to the edge of the gill-covers three inches. Under jaw a little the longest; eye large; upper maxillary bone terminal, the snout receding from it backward, contrary to the form of the Whiting, in which the upper jaw is under a projection. The general shape of the body more slender than in the Whiting, but that this did not proceed from emaciation is shown by the roundness of the back, which was plump. Prominent teeth in the jaws, and at the roof of the mouth a pair of prominent, sharp, incurved teeth. Lateral line straight, passing along near the back; another line along the middle of the side, formed by the meeting of the muscles; the body ending more slender at the caudal fin. The first dorsal begins over the posterior third of the pectoral; second dorsal like the first in shape and elevation, both being triangular; between them a space about equal to their individual breadths; nearly twice this breadth between the second and third dorsal fins; the beginning of the third dorsal slightly anterior to the second anal fin; caudal fin shaped as in the Whiting, but less wide; the pectoral ends opposite the middle of the first dorsal; ventrals small and slender, rather high on the side, and much like those of the Pollack; the longest ray seven eighths of an inch in length. From the vent to the first anal fin a quarter of an inch; first anal long, widest in the middle; second anal longer than the third dorsal, both ending close to the caudal fin. Colour on the back brown, sides much lighter, belly white; eye yellow, lighter yellow on the gill-covers. A dark spot on the upper border of the origin of the dorsal fin; along the base of the anal fins a broad white band, but not at the margin, and this remained unaltered after the brilliancy of all besides had greatly faded. The distinctions between this fish and the Common Whiting are obvious in the more slender shape, the jaws, teeth in the palate, lateral line, fins, and colour. The number of rays in the fins were,—in the dorsals thirteen, twelve, and twenty-two; and thirty-five and twenty-five; pectoral twenty; ventral six.

POLLACK.

WHITING POLLACK.

Asellus huitingo-pollachius, a Whiting Pollack,	WILLOUGHBY; p. 167.
Gade pollack,	LACEPEDE. RISSO.
Gadus pollachius,	LINNÆUS. BLOCH; pl. 68.
" "	DONOVAN; pl. 7.
Merlangus pollachius,	FLEMING; Br. Animals, p. 195.
" "	JENYNS; Manual, p. 446.
" "	YARRELL; Br. Fishes, vol. ii, p. 253.
Gadus pollachius,	GUNTHER; Cat. Br. M., vol. iv, p. 338.

THE Pollack is one of our commonest fishes, and is found on all our coasts where the nature of the ground is suitable to its habits; but it becomes more scarce in the extreme north of Scotland. It is also one of the few species of this family that is met with in the Mediterranean, but according to Risso it is not abundant in that sea.

Its haunts are at no great distance from land, and it prefers to keep amidst rough and rocky ground, where it lies in wait on that side against which the tide happens to be flowing, and, advancing from which, it is prepared to seize whatever prey may come within its sight. It wanders from one station to another, and if a considerable number are found together, it is that they are drawn together by the attraction of prey, the motions of which they follow with eagerness. They then swim at a good distance from the bottom, and rise or fall as they find occasion, by which it happens that they become entangled in the trammel nets which are set near the ground; and again it is common to see them in rapid action close to

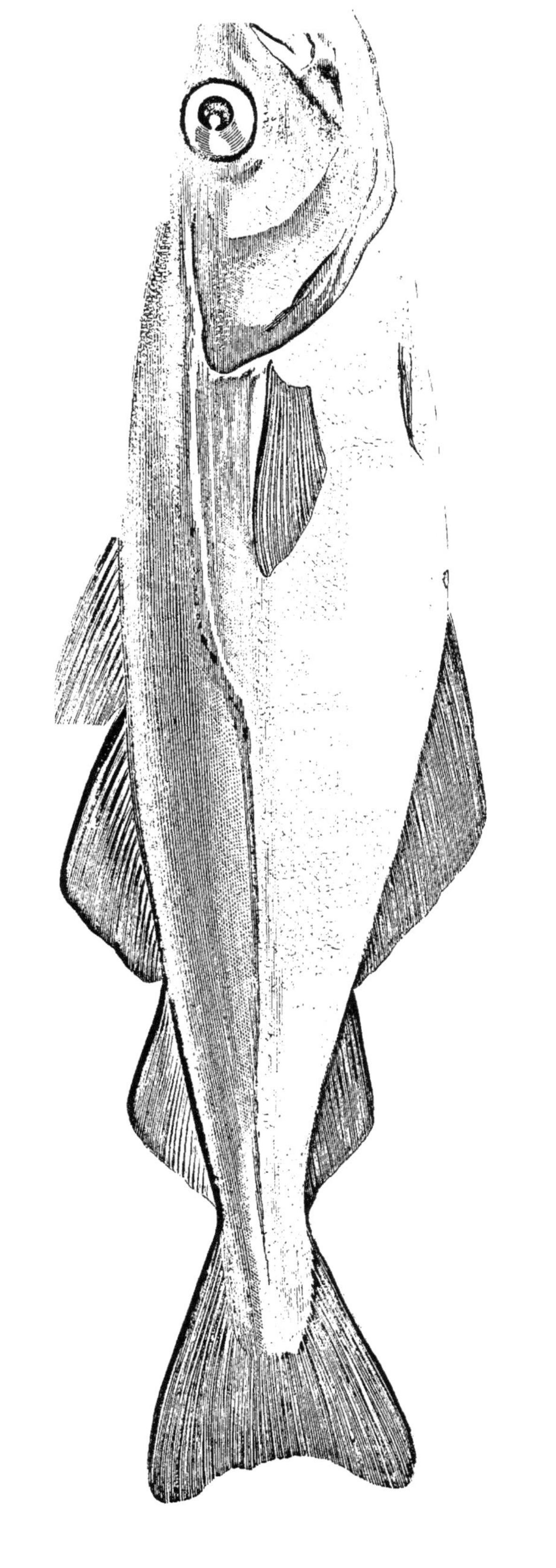

the surface, in pursuit of the younger fishes which are frisking in unsuspecting security near the shore, when the sea itself appears to be asleep. It is then that schools of the smaller Launce, and other fishes newly born, which swim near the surface, are broken in upon by repeated plunges, which disturb the quiet to a considerable distance; and as the affrighted companies again collect together, again and again there is a plunge, until the pursuers are glutted, or the pursued have reached a place of safety.

It appears that even young Pollaeks are persevering persecutors of fishes which are smaller or more helpless than themselves, of which the following incident. communicated by Mr. Peach, is an amusing instance:—A small Whiting was observed to have taken shelter within the hollow of a medusa, *(Cyanea aurita,)* a circumstance of common occurrence with very young fishes of several kinds; but in doing this the action was observed by a young Pollack of about five inches in length, which immediately began an attack. The little Whiting easily evaded these attacks by dodging round its friend; but the pursuer was soon joined by another of its own kind, and both of them united in the same work. For a time both of them were baffled; but an unlucky movement drove the pursued one from its shelter, and a severe chase immediately took place. Several additional Pollacks joined in this chase like a pack of hounds, and in terror the Whiting rushed to the surface. The pursuit was doubtful; but at last the hunted one became exhausted, and lay as if dead, so as to be drifted along with the tide. After a time, however, animation was recovered, and the little Whiting again found refuge within the cavity of the medusa. This movement, however, was presently discerned by the congregated Pollacks, which allowed it little respite. They soon drove it into open water, and after a short chase it fell a victim to their violence, and this too without their proceeding to feed on the carcase. So eager were these Pollacks in the pursuit, that when stones were thrown to drive them away they shewed no alarm, although at other times a single stone would have struck them with terror.

Dr. Fleming says they are sometimes caught by employing a white feather as a bait—we must suppose at the surface;

but the usual method of fishing for them is in the manner called whiffing, by using a length of line which is not weighed down with a sinker, and is towed after a moving boat. The bait is made, both by the setting on and action, to imitate a living object, and the fisherman manages two of these lines by alternate motion of his arms, while another rows the boat. They are the half-grown fishes which are thus caught, and the larger fall victims to the ordinary line at anchor, chiefly in the autumn, at which time these fishes abound in considerable numbers, and are in their highest perfection; in which for the table they are little inferior to the most esteemed of the family. Like the Whiting, however, they do not vary greatly through the year; but the best, as well in size as quality, are those which are caught at the West of the Land's End, between that point and the Scilly Islands; a district in which others of the Gadoid family are also found to reach an amount of perfection that is not equalled elsewhere.

The Pollack spawns about the end of the year, and the young, of small size, are seen in harbours, and on the borders of shallow rocks, moving about with a slow motion, and ready to take a bait as it comes in their way.

It is often salted and dried by fishermen for their own use, but in this state they do not usually form an article of trade, although fully equal to some that meet a ready sale.

The form is compressed, moderately lengthened. The usual weight from twelve to fifteen pounds, and very rarely exceeding twenty; but our description and figure were obtained from an example that weighed twenty-four pounds. Under jaw protruding beyond the upper; numerous small teeth in both, and strong teeth in the palate. Eye moderately large. Scales on the body small and well retained. The greatest depth at the vent, which is opposite the middle of the first dorsal fin. Lateral line with a curve, which sinks a little beyond the end of the pectoral fin, and from thence straight to the tail. The first anal fin narrow, and sloping in its outline, as, in a less degree, is the second anal also. Pectoral fin narrow; ventral small and further back than in most of this family. Tail a little incurved. The colour of the back and fins is a dark brown or olive, sometimes with a tinge of green; sides often obscurely mottled with brown or yellow; belly obscurely white. In the younger condition it is occasionally found with the sides and belly yellow or a bright orange, a colour they are supposed to assume from living in the shelter of rocks clothed with oreweed. The first dorsal fin has thirteen rays, second nineteen, and third seventeen; first anal with thirty-one, second nineteen, as has also the pectoral; ventral five; caudal perhaps thirty-four.

COALFISH.

RAUNING POLLACK.

Asellus niger,	Willoughby; Table L., p. 168.
Gadus carbonarius.	Linnæus.
Gade colin.	Lacepede. Risso.
Gadus carbonarius,	Donovan; pl. 13. But his figure is faulty, not only in the colour, but in placing the dorsal and anal fins too far separate from each other.
Merlangus carbonarius,	Fleming; Br. Animals, p. 196.
" "	Jenyns; Manual, p. 466.
" "	Yarrell; Br. Fishes, vol. ii, p. 250.
Gadus virens (?)	Gunther; Cat. Br. M., vol. iv, p. 339.

When we seek for information from different sources on the natural history of fishes, we are liable to be misled by finding that different kinds are sometimes called by the same name, and more frequently that one species shall bear a multiplicity of names, even in districts not very distant from each other. In the history of our "Fishes of the British Islands" we have found it generally convenient to omit all reference to these local designations, as having little meaning attached to them, and which we should be well pleased to find discarded from the memory. But for once we depart from our rule that we may record an instance of these variations of denomination; and it seems the more appropriate in this instance, as it forms almost an integral portion of the history of the Coalfish, which is thus more diversely characterized than any other with which we are acquainted. In Ireland it bears a different name according to its stage of growth: the very young being known as Gilpins, from which they grow to be Blockan and Greylords; and when of full size they are Glashan, or Glossan, and Glassin. Moulrush and Black Pollack are other names, with Glassock; Billets and Billards in Yorkshire: Sey Pollack, Podley, Sillock, Cooth, Pittock, Sethe, Colmey, Harbin, Coalsey, Cudden, and Green Cod. Willoughby says it is called Rawlin Pollack in Cornwall, which is grounded in nothing more than error of

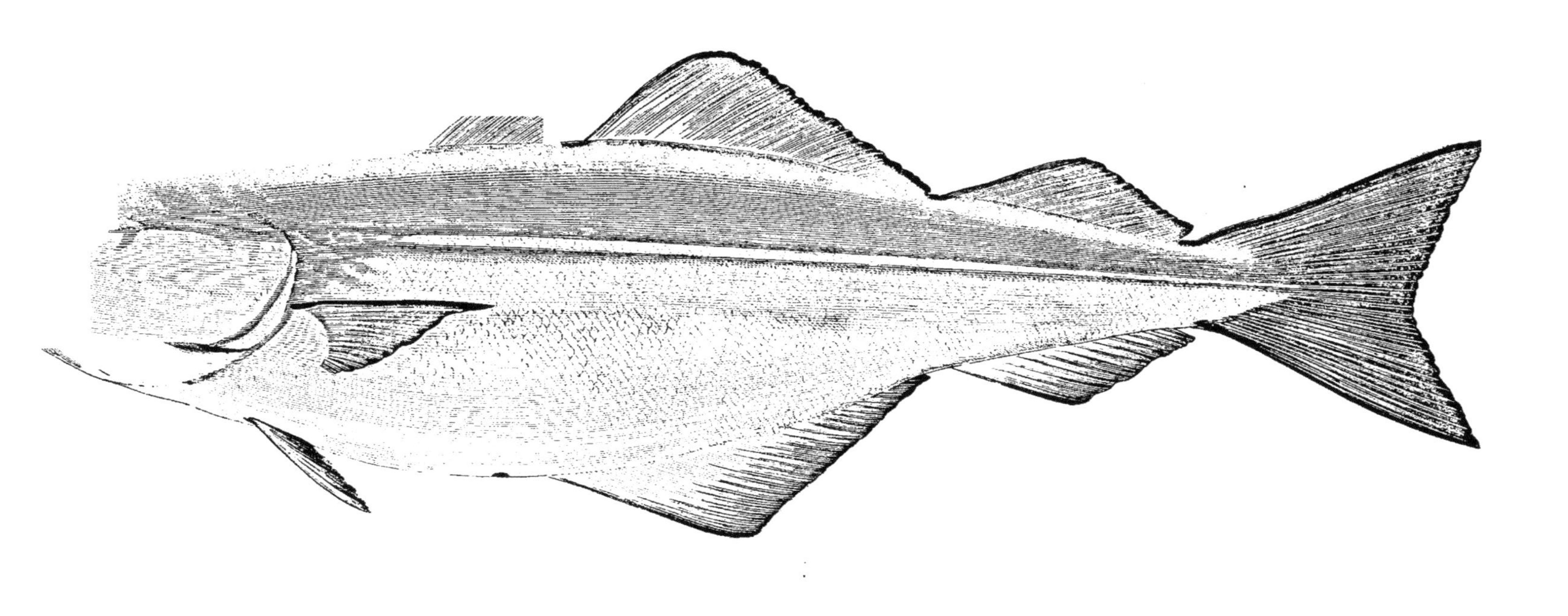

COALFISH.

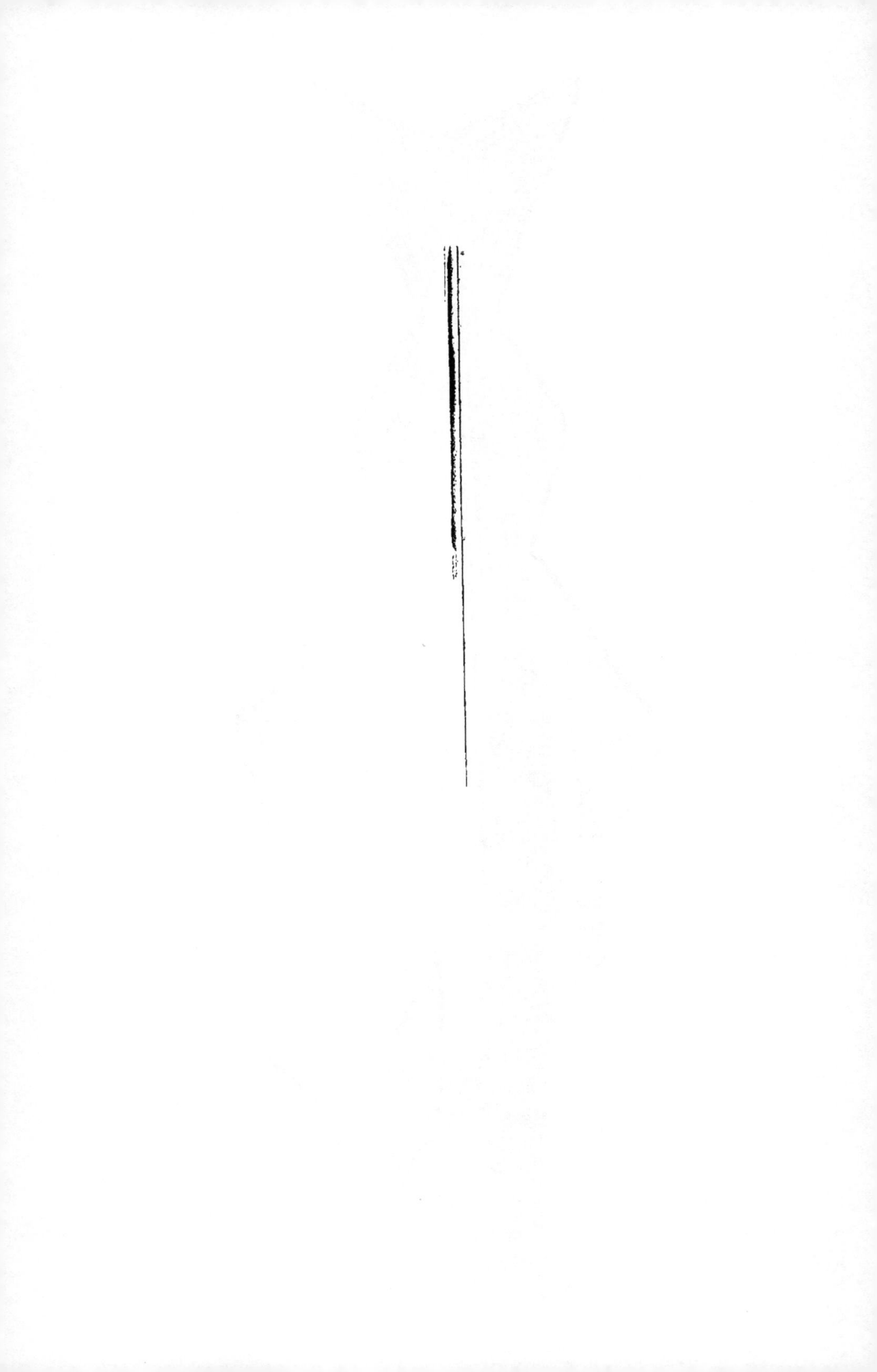

pronunciation; but Lacepede has carried the mistake a little further, in supposing it to bear the name of Raw Pollack. This word, Pollack, appears to have been taken from the northern nations, since it is the name employed by the people of Scandinavia.

The name of Coalfish carries with it its own signification, but it has not been always understood; and the framer of an Act of Parliament, (15th. Charles II, C. 7,) not appearing to know what relationship there could be between this fish and the mineral, but supposing perhaps that the gull was a bird which had some connection with the sea, affixed to it also the name of Gullfish. It further becomes a question whether even another name should not be added to this lengthened list; for in an Act of Parliament, called the Statute of Herrings, (the 31st. of Edward III, A.D. 1357,) there are the names of three fishes associated together, the taking and sale of which were thought worthy of being regulated by law, and that too in an arbitrary manner. These were Lob, Ling, and Cod; of the two latter of which there cannot be a doubt; but the former is more obscure. In Wright's "Dictionary of Obsolete Words" the word is said to mean unwieldy—a lump: the proper name of the fish would appear to be Lobkeling. According to a Cambridge manuscript,

"Lobkeling catcheth spirling—
So stroyeth more men the lesse."

Another dictionary says that lob means lazy—lumpish. It may mean the Coalfish or Hake.

It is common and abundant on all the coasts of the British Islands, but the numbers are much the greatest in the north, and its range appears to extend in that direction as far as enterprise has yet reached, if indeed, in this last instance, the species is not different. It is known in North America, and Risso classes it also with the fishes of the Mediterranean, but he says it is rare at Nice, where he wrote, and he knows it only as taken in June.

It is eminently a ravenous fish, and its Cornish name is characteristic of that propensity, the expression rauning being the ancient, and in some places the present pronunciation of the word ravening or ravenous. It snatches at a bait with headlong eagerness; and as, especially in autumn, they often swim close to the surface in considerable numbers, when a fisherman has the good chance to fall in with a company, he will generally succeed in securing a large number, if not the whole; so that I have known four men with two boats (two men in each boat) secure twenty-four hundredweight with lines in a very few hours. The size of each fish ran at twenty-five pounds with great regularity. It seems uncertain what may be the object of their thus collecting together, but they are swift in their motions, and sport with the same energy that they devour. At other times, besides these gatherings together, the Coalfish associates but little with its fellows.

Although this fish is not seen at fashionable tables, it will in its season bear some comparison with a portion of the same family that are. It is at least valued by those who are not fastidious in their choice; and, as it takes a preparation with salt favourably, large quantities are cured in the north for exportation. Mr. Edmonson, in his "View of the Zetland Islands," informs us, that besides the quantities that are used fresh, about fifty tons are exported every year. In the west of England they are chiefly kept for the use of the fisherman's family.

They spawn in the spring, and we are informed that in the islands north of Scotland in the summer the young abound, and are angled for from the rocks, and serve a good purpose in the support of the poorer classes. They are much preyed on by other fish.

The length of the Coalfish is about three feet, with the weight perhaps of thirty pounds, and a shape well fitted for active exertion. Head pointed, a little flattened above; under jaw longest, but the proportion less than in the Pollack; teeth in both, and a few in the palate. Body plump, compressed, more slender towards the tail; scales small; lateral line straight, whitish, conspicuous. Vent opposite the division between the first and second dorsal fins. Pectoral fin pointed; dorsal and anal fins rather more angular than in the others of this genus; tail concave. Colour black on the back and dorsal fins, lighter below. The first dorsal fin has thirteen rays, second twenty-one rays, third nineteen; the first anal twenty-four, second fourteen; pectoral twenty; ventral five; caudal thirty-four.

GREEN POLLACK.

GREEN POLLACK.

FROM a distant date a fish called the Green Pollack has occupied a place in the list of British natural history, and it has a station in the system of Linnæus, with the name of *Gadus virens*. Other references are:—

Asellus virescens,	WILLOUGHBY; p. 173, table L. M. N. 1. But he had never seen the fish, and supposes it to be the young condition of the Pollack, which his own figure might have taught him it was not.
Gade sey,	LACEPEDE. RISSO.
Merlangus virens,	FLEMING; Br. Animals, p. 195.
" "	JENYNS; Manual, p. 447.
" "	YARRELL; Br. Fishes, vol. ii, p. 256.

WE believe that a notice of the Green Pollack, in the supposition that it is a distinct species, was communicated to Pennant by Sir John Collum, who obtained it in Devonshire; and if so the fish must be the same as that with which we are acquainted, and as is represented by Mr. Yarrell, in which case it certainly is as we have described it. But our fish does not closely answer to the figure given by Fries and Eckstrom. I possess no other than the first edition of Pennant's work, in which there is no account of the Green Pollack.

This fish is common, and at times abundant, even in considerable schools, which are sometimes seen in harbours, or the close neighbourhood of the shore. But without hesitation I express the opinion that it is only the young form of the Coalfish or Rauning Pollack, of which the gradations may be traced in all stages of its growth from five or six inches in length (which is the usual size of what is termed the Green Pollack) to about a foot, at which size the undoubted marks of the Coalfish shew themselves. If we could suppose these fishes to be of different species, the distinctions between them would be, that in the Green Pollack the jaws appear equal, and the lateral line without the white appearance; but when the other characters of the Coalfish are seen the jaws still remain with little difference of length, and the small amount of final increase is almost imperceptibly given. The green colour of the back, and yellow hue on the sides give place to a darker tint when the fish has reached the length of about ten inches, and it becomes still blacker with the increase of age.

The name of Sey Pollack, by which this fish or the Coalfish is known in most northern districts of the British Island, appears to be of Scandinavian origin, and, with a distinctive adjunct, is applied by Nilsson to several species. It is to be observed further, that something more than doubt is felt by eminent naturalists, whether the Coalfish is to be found in regions so close to the polar ocean as has been supposed. On the contrary, as already intimated, it has been suggested that this more northern fish is a distinct species.

Deformity in fish is not of rare occurrence, but in one instance a Coalfish of large size was met with that seems to require particular notice. The upper jaw was shortened in such a manner as to give the head a peculiar appearance. The body was depressed out of the regularly straight shape. But the most remarkable deviations from what is usual was in the fins, of which the first dorsal was more narrow, lofty, and sharper, having its origin nearer to the head; between it and the second dorsal was a vacant space, and a much larger space between the second dorsal and the third, both of them unnatural. Three anal fins; the first an irregular triangle, second very narrow, and a considerable space in a curve between it and the third, which also was unnatural. The vent was much behind its usual situation. This fish appeared to be thin and ill fed.

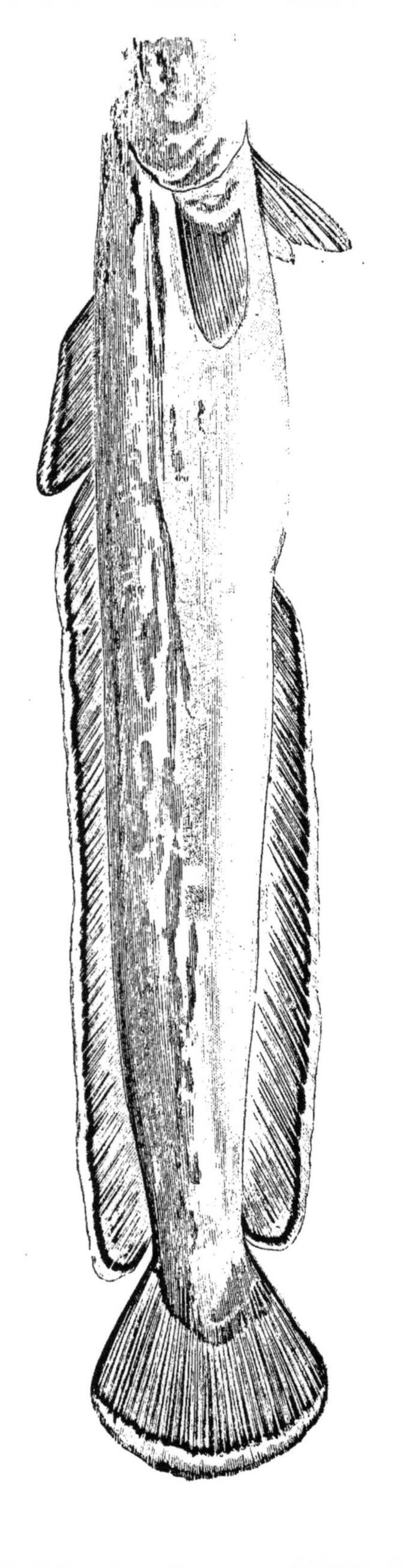

LOTA.

With the general characters of the Gadoid family, the body is more lengthened; a barb at the lower jaw; two fins on the back, the second and anal long.

LING.

Ling,	Willoughby; p. 175, Table L. 2, N. 2.
Gadus molva,	Linnæus, Bloch; pl. 69. Donovan; pl. 102.
Gade molve,	Lacepede. Risso.
Molva vulgaris,	Fleming; Br. Animals, p. 192.
Lota molva,	Jenyns; Manual, p. 448.
" "	Yarrell; Br. Fishes, vol. ii, p. 264.

The Ling is one of the commonest fishes in all parts of the British Islands, and is nearly in the same estimation for the table as the Cod,—not, indeed, when it is fresh, but salted and dried it is preferred; and especially when caught in that space of sea which lies between the Scilly Islands and Land's End, where we suppose the Ling to be found of larger size and higher perfection than in any other part of England. Its cherished resorts are on the rocky borders of the wider valleys of the sea, at some distance from land; but, as the time approaches for shedding the roe, they assemble in considerable numbers somewhat nearer the land; and at this time they are particularly sought after by the fishermen, who employ for taking them the ordinary lines from the boat, which are weighed down with leads proportionally heavy according to the force of the tide. Another and more successful method is with the long line, or bultey, which usually is formed of several hundreds of hooks arranged along a principal line, such as has been already described, and, when stretched out across the tide, extending even to a few miles in length. It is an economical method of proceeding in some districts of the north of Ireland, as we learn from Mr. Brabazon's account of the fisheries of that country, that several individuals will join together in providing the proper length of line, in which each adventurer is the proprietor of so much of the extent of it as he has provided, and of which he takes the produce, to the exclusion of others; but, we suppose, with some reserve of common interest.

The numbers thus taken, of all sorts of fish, are sometimes very great; but where a ready sale is not obtained, the greater portion is preserved in the usual way, and dried for exportation. The different parts of Italy receive a large proportion of these salted fish. But the consumption of salted Ling, which even now is considerable at home, was formerly of very large amount, and it was even an ordinary dish at royal and noble tables. In the Rutland Papers, printed for the Camden Society, we are told that on the visit of the Emperor Charles the Fifth to London, in the reign of Henry the Eighth, salted Ling was among the principal matters provided for the entertainment of the guests; and in the directions given to the Lord Mayor to guide him in his preparations, he is ordered,—"Item, to assigne two fysshemonngers for provision of lynges to be redy waterd." Although the taste appears to have declined in the reign of the first James, the practice seems to have maintained its ground; for, among the pieces of merriment of this king, he is said to have professed that if his royal brother of the lower regions should be pleased to visit him, his dinner should consist of a pole of Ling and mustard, with another equal favourite of his, a pipe of tobacco for digestion.

According to Fuller, in his "Worthies of England," the extent of the adventure was equal to the value set on the fish. Referring to the mischief wrought by the civil war, he says:—"We are sensible of the decay of so many towns on our north-east sea, Hartlepool, Whitebay, Bridlington, Scarborough—and generall all from Newcastle to Harewich,—which formerly set out yearly, (as I am informed) two hundred ships and upwards, imployed in the fisheries, but chiefly for the taking of Ling, that noble fish." That it formed an ordinary article in the provision for families in the winter appears from the advice which Tusser gives in regard to its safe keeping. In his husbandry for winter he recommends—

> "Both salt-fish and Ling, if any ye have,
> Through shifting and drying from rotting to save;
> Lest winter with moistness do make it relent,
> And put it to hazard before it be spent."

Oil from the liver of this is probably of the same value as a medicine with that of the Cod; and it was formerly much used, with that of several other fishes, to light the fisherman's lamp, instead of a candle. The air-bladders also, or sounds, are much valued as a delicate food, although they fall greatly below those of the Cod.

The season when the best captures are made varies on different parts of the coast. In the west of England it is in January and February; but, according to Dr. Edmonson, in Zetland this fishery begins on the 20th. of May, and ends on the 12th. day of August. Mr. Brabazon says that in the north of Ireland it is in March, and that this fish spawns there towards the end of May. On the re-discovery of the abundant fishing-ground at Rockall, while the Cod was the principal object, the Ling was not found to be deficient.

The Ling is an eager feeder, and yet it displays some degree of choice in the selection of its food, among which a living prey is preferred, although it will readily swallow a piece of herring, pilchard, or a slice of two or three kinds of cuttle. Pieces of the conger are used as bait in some places, but while crabs and the lobster kind find no acceptance; (for in a large number that were examined, none of these animals were found, although the stomachs of the Cod caught in the same ground contained them in plenty.) Lacepede remarks that this fish shews a decided preference for the Plaise; and, in confirmation of this, from one Ling seven of the last-named species, which varied in length from six to ten inches, have been taken. Skulpins also are sometimes found, and even a Rough Hound of considerable size.

That even severe injury to the stomach does not deprive this fish of its craving appetite is shewn by the fact that when a large hook had gone down through the gullet with its shank foremost, and thus had penetrated through the side of that organ, a turn of the hook had brought the point again through the substance to pass through the bottom in the same manner, and again through the opposite side, so that, by drawing the whole together, it left only a small cavity free, in which condition the line was bitten through, yet the fish escaped only to swallow another bait, with which it was caught.

As might be supposed from the multitudes obtained this is a prolific fish, and the roe of a large one has been known to weigh eleven pounds. When about half developed, the lobes of roe, as well as those of the Cod and Whiting, when fried or roasted are thought a delicious dish.

The Ling, as its name signifies, is of a more lengthened form than any other of this family, and the shape, which is generally uniform, only becomes somewhat more slender and compressed towards the tail. Head flattened on the top, slightly compressed at the sides; upper jaw reaching a little beyond the lower; numerous teeth in each; palate encompassed with teeth of different lengths. Eye moderate; a barb at the under jaw, and I have seen an example where there were two barbs. Vent nearer the tail than the origin of the second dorsal fin. Scales scarcely perceptible; lateral line descending gradually beyond the beginning of the second dorsal, and from thence straight. Of the two dorsal fins the first begins above the middle of the pectoral, and rises but little above the line of the second, the origin of which is close to the termination of the former. The second dorsal and anal run evenly close to the tail, where they become a little expanded. Tail round, as also is the pectoral fin, and in some degree also the ventral. Colour of the back light or yellowish brown, mottled with dull yellow at the sides, belly white. Dorsal and anal fins and the tail edged with white. A Ling five feet and a half long has weighed about seventy pounds; but I have been informed of an example caught near the Scilly Islands, which weighed one hundred and twenty-four pounds. It was offered to the gentleman who informed me of the circumstance, at the price of a shilling, but this was before the opening of a railroad in that direction.

The first dorsal fin has fourteen rays, the second sixty-six; anal sixty-two; pectoral twenty; ventral five.

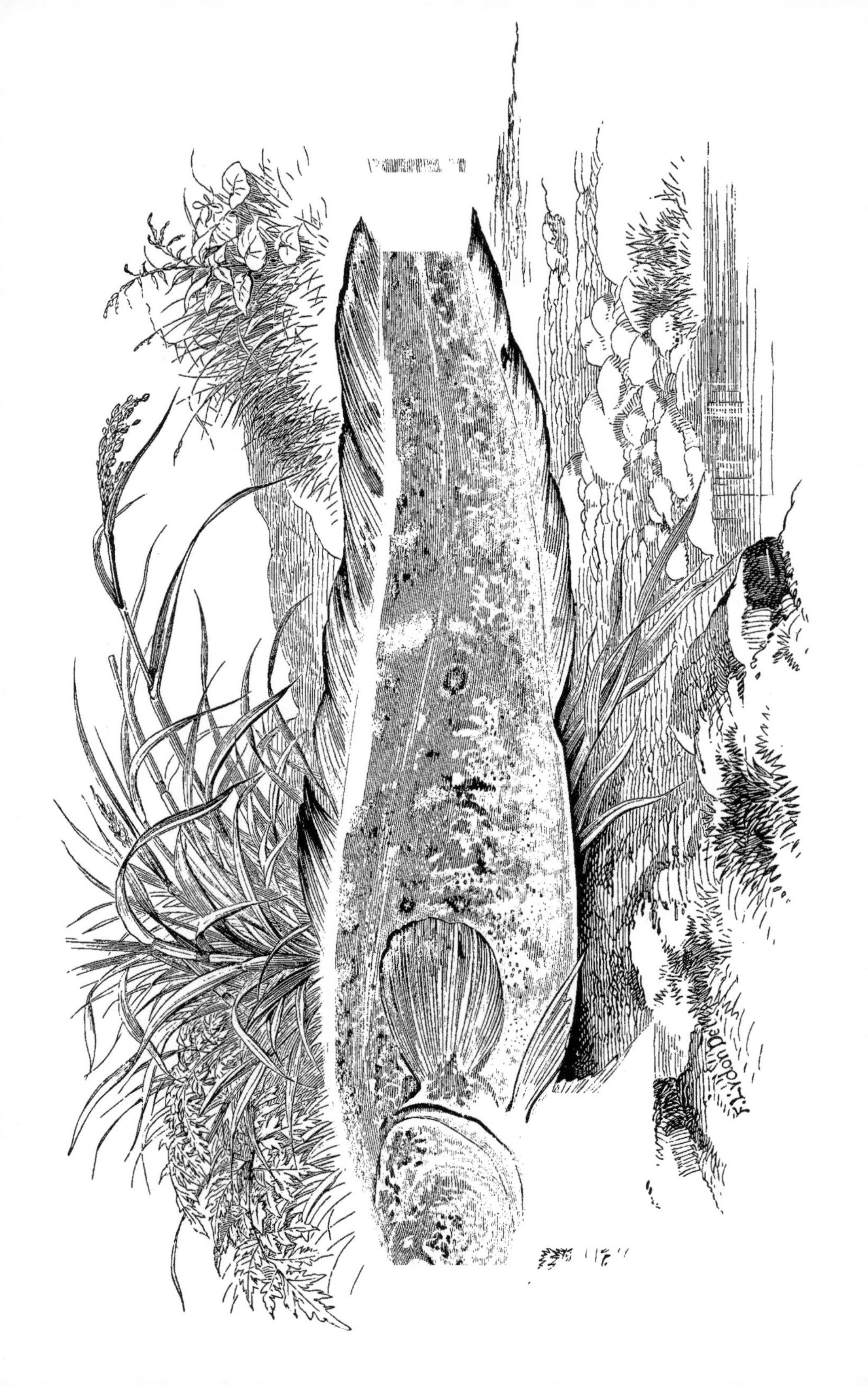

BURBOLT.

BURBOT. EELPOUT.

Lota gallorum,	Jonston; Table 29, f. 10.
Mustela fluviatilis,	Willoughby; p. 125, table H 4.
Gadus lota,	Linnæus. Bloch; pl. 70.
" "	Donovan; pl. 92.
Garde lote,	Lacepede.
Molva lota,	Fleming; Br. Animals, p. 192.
Lota vulgaris,	Jenyns; Manual, p. 448.
" "	Yarrell; Br. Fishes, vol. ii, p. 267.

The Burbolt is the only one of the extensive family of the Codfishes which has its residence in fresh water, where it is distinguished by exhibiting some of the manners of the eel, by which it has obtained in some places the name of Eelpout; but the habits in which these last-named fishes agree are so few as to shew little more than a distant analogy, while the difference of shape is a proof of the little foundation there is for Lacepede's opinion that there is a real affinity between them.

In England at least this fish is thinly distributed; which circumstance may arise from its finding congenial haunts only in deep and slowly-moving streams; but it may also have been caused by its having possibly been an imported fish; which the excellency of its flesh, and its power of sustaining with proper care long absence from its native element, render not improbable. None are found in Scotland, although from its hardy nature the cold of that country can offer no difficulty. But it is found in the rivers of Yorkshire and Durham, Norfolk, Lincolnshire, and Cambridgeshire; and also the Tarne, but not further westward, nor in Ireland; to the lakes or ponds of which it might be easily conveyed with much profit to the public. A considerable degree of cold, at least in winter, appears to be natural to it; and hence it is not only common in Sweden, and other parts of the north of Europe, but also in Siberia and other portions of the north of Asia, as well as in India.

Besides the names which we have already assigned to it, this fish is also locally known as the Coney fish, from an opinion formerly held that the creature called the Coney in the Sacred Scriptures,—the Arkeeko of Bruce, is the same with our common rabbit; and this fish so far imitates the animal of the land as to pass much of its time, and seek its shelter in holes and overhanging banks of the rivers it frequents. These are its hiding places by day, and from them it proceeds to seek its food in the evening and night; and it is at these times that the chief success is obtained by fishing for it: the method being by lines laid especially for the purpose. We are informed that forty have been taken in the River Trent by one fisherman in a single night; and indeed with a little skill in finding their haunts, there seems to be little difficulty in securing them, for they possess the common character of their tribe in being very voracious; and Sir John Franklin, in his first voyage to the far north of the American continent, where he discovered this fish, and which there bears the name of Methy, observes of it that it will prey on fish that are large enough to swell out its body to almost twice its natural size. To the commendation bestowed on the flesh of the Burbolt, the liver is also pronounced a great delicacy; but it is added that the roe is almost poisonous. This roe is produced in great abundance.

It is said that the Burbolt is found of larger size on the continent, and particularly in the Lake of Geneva, than with us; but we have a record of an example which weighed six pounds, and Pennant mentions one which amounted to eight pounds. Lloyd, in his Scandinavian adventures, mentions a Burbolt that weighed twenty pounds. A more common size is two or three pounds. The example selected for description measured in length thirteen inches and a half. Head depressed, wide, sloping from above the gill-covers to the mouth; upper jaw a little overlapping the lower; snout rounded; mouth rather capacious, tongue large, teeth numerous, in a bed round the jaws, and a wide circle round the palate. Barb on the lower jaw slender; eye moderate; body round and stout, with a depression at the beginning of the back; afterward tapering and compressed to the tail, but not proportionally as long as the Ling. Lateral line not well made out; scales numerous, round, appearing as if each formed a depressed cup: they run over the cheeks, and are very small forward to the eyes. Vent very little before the middle of the body, exclusive of the tail fin. Pectoral fins large and round, in which were imperfectly counted twenty-four rays. The first dorsal fin begins at the hindmost border of the pectoral, level along its border; the second begins close to the first, and ends by being joined to the border of the caudal; the anal runs parallel with it, and both are a little spread out at their termination. Ventrals a little before the pectorals, wide apart, with seven rays, of which the outermost is an inch and a half long, ending in a point, the others becoming gradually shorter; the tail almost lancet-shaped. The general colour is a rich brown, darker on the back and borders of the fins, with browner variegations; on the head and cheeks mottlings of dark brown; two or three dark spots on the second dorsal and upper part of the body behind; base of the pectoral yellow, white below.

BROSMIUS.

THIS genus is marked by having only a single fin on the back, a lengthened body, and a barb at the chin. It is therefore an aberrant form of the true gadoid fishes; but it agrees with them in all points, except in the absence of a first dorsal fin.

TORSK.

TUSK.

Gadus brosme,	LACEPEDE. DONOVAN; pl. 70.
Brosmus vulgaris,	FLEMING; British Animals, p. 194.
" "	JENYNS; Manual, p. 452.
" "	YARRELL; British Fishes, vol. ii, p. 285.

TURTON's edition of Linnæus represents *Gadus brosme* and *G. scoticus* to be distinct species, the former being perhaps the *Brosmius lub* of modern writers. The fish itself was unknown to Artedi and Linnæus. In the Scandinavian languages the word Torsk is applied to the species of Codfishes in general, as distinguished from the Pollacks and Lings; but in the northern portions of the British Islands it has become the name of a fish not belonging to that section of gadoid fishes to which the people of the north had confined it.

The fish so named in England is a native of the northern seas, and is met with in abundance in the neighbourhood of the Orkney and Zetland Islands, where it is the object of a fishery of considerable local importance. On the newly re-discovered ground at Rockall it exists in common with the Cod and Ling, but it becomes more rare as we come southward; and, although it is sometimes caught in the Moray Firth, there is no instance on record of its being met with in

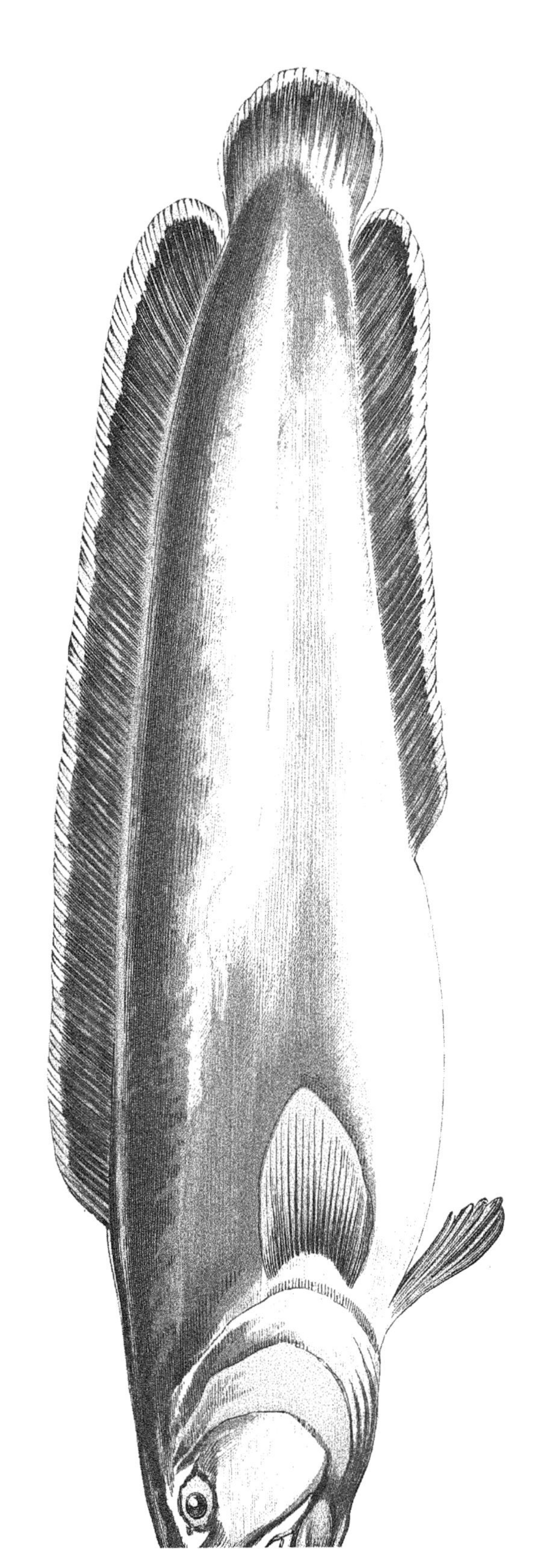

England. As it makes some approach to the Ling in shape, so also its habits appear to be not greatly unlike, its resort being near the bottom, where it is eager in searching after food. It is also in estimation for the table, and large numbers are salted in the northern islands for exportation. It is classed among the fishes of Sweden, but it appears that its distribution in the higher regions of the north is within such moderate limits as that it is not met with at stations which are inhabited by species which are also common on the southern coasts of England and Ireland. It is said to be known on the south and west of Greenland, but not on the colder eastward coasts, although it has been doubted whether this more northern fish be not a different but closely allied species.

The Torsk prefers the deeper water of the ocean, and usually the rougher ground; but it comes nearer the coast at the time of spawning, which is in January and February; and we are informed that when in the shallower water it becomes liable to receive injury from the storms that then prevail, so that great numbers of them are sometimes thrown dead on the shore. The stomach of this fish when caught is commonly found empty, which is to be accounted for from the like cause by which that organ in the Cod, Ling, and several others of the family is often seen to be everted. It appears to proceed from the strong influence of terror, by which, in a less degree, it also happens that the contents of the stomach in these fishes are discharged when the fisherman is drawing them from their depths; or, as in the case of the Hake, when hooked high in the water, the multifarious contents are ejected only after it is taken into the boat.

The example selected for description, for which I am indebted to the Rev. Walter Gregor, of Macduff, was in length sixteen inches; the general form much like that of the Ling, but proportionally stouter; the body thick, more compressed behind; head and origin of the body broad. The body deepest opposite the origin of the dorsal fin. Snout slightly over the upper jaw; when closed both jaws equal; gape capacious; teeth incurved and strong; a barb at the lower jaw. A depression runs backward from the head, in which is placed the single dorsal fin. This fin begins opposite half the length of the pectoral; anal at the middle of the body, and both pass on and end close to the origin of the caudal fin, which is small and round; the dorsal and anal a little expanded towards their termination. Neither the scales nor lateral line are to be seen when the fish is fresh from the sea; the scales very small; the line begins high behind the head, and sinks down at half its course. Pectoral fin oval; ventrals rather long. Colour dull yellow, darker on the back; pectorals yellow; dorsal and anal fins dark, with a border of faint yellow; the tail also with a border; ventrals dark. With some doubt the fin rays were counted,—dorsal ninety-seven, anal seventy, pectoral twenty-two, caudal twenty-eight.

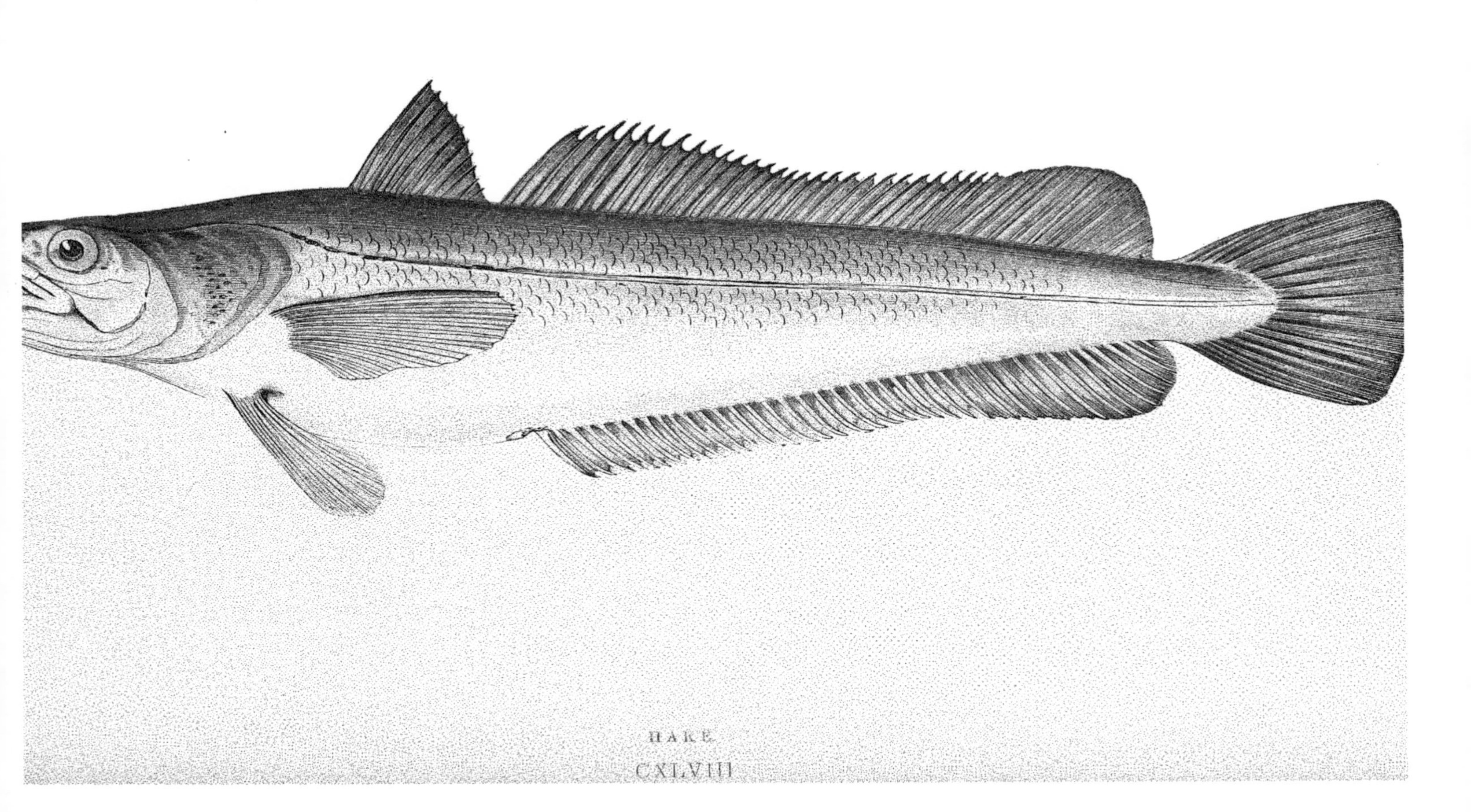

HAKE.

CXLVIII

MERLUCIUS.

THE generic character is, that there are two dorsal fins and a single anal fin; but there is no barb at the chin.

HAKE.

Asellus sive merlucius,	JONSTON; Table 1. f. 3.
" " "	WILLOUGHBY: p. 174.
Gadus merlucius,	LINNÆUS; but he speaks of it as having a barb.
" "	BLOCH; pl. 164. DONOVAN; pl. 28.
Merlucius vulgaris,	FLEMING; British Animals, p. 195.
Gade merlus,	LACEPEDE.
Merlucius vulgaris,	JENYNS; Manual. p. 447.
" "	YARRELL; British Fishes, vol. ii, p. 258.
" "	GUNTHER; Cat. Br. Museum, vol. iv.

THE Hake is one of our commonest fishes round the coasts of the British Islands, but it abounds in the greatest numbers in the south and west of England and Ireland. Bellamy, in his account of the fishes met with in the south of Devonshire, says that sixty thousand were brought by trawlers into Plymouth in the months of December and January; and I have been informed that forty thousand were landed in Mount's Bay in one day; and on another occasion eleven hundred were taken by one boat in two nights,—the evening or night being the most successful time of fishing for them. These large assemblies however are not appropriate to the usual habits of this fish, and their gathering together no more proceeds from a love of union than does that of a cry of wolves when hunting their prey. They watch the movements of smaller fishes, and devour voraciously the pilchards and herrings which throng the coasts, in feeding on which it is only when gorged to excess that their craving is satisfied. When a school of Pilchards is enclosed within a sean, it will commonly happen that several Hakes are cooped up with them; and when the tucking of the sean is in progress, for the purpose of taking up the imprisoned fish, the Hakes are often found so filled with the smaller fish as to be utterly helpless. Seventeen Pilchards have been found in the stomach of a Hake on this occasion. As even the fishes of the sea are conscious of an instinctive dread of danger, fishermen affirm that the smaller tribes display as significant a fear of the approach of the Hake and Ling as of a Shark, and they quit the station when these enemies appear.

On ordinary occasions Hakes swim at a considerable depth in the water, and shew themselves ready to seize whatever object may chance to offer itself to their notice; but, contrary to the more usual habits of their family, they appear to lose their appetite at the time when they are about to shed their spawn, the season for which is the early months of the year, although this is liable to variation, as indeed is the case with most fishes, so that in the cold season of 1837 the spawning of Hakes was not accomplished until August. At this time their presence on the coast is signified only by the numbers taken with trawls, while very few fall to the lot of those who fish with hook and line. When caught with the latter at a considerable depth this fish ejects the contents of its stomach before it reaches the surface, but when the hook has been swallowed at a higher station in the water this ejection does not take place until the captive is drawn on board; and there can be little doubt that terror is the exciting cause in every instance; but the fact itself goes far to explain how it happens that many fishes which beyond doubt are eager feeders, are generally found when caught to have their stomachs altogether empty.

There is reason to believe that in early times the Hake was far more highly valued for the table than we now find it to be; but it is probable that it then filled the place now occupied by the Cod of Newfoundland, as well as of being an accustomed dish for the fast in Lent, and other usual days of abstinence enforced by law. In the fifteenth century salted Hakes formed an important part of the trade of Ireland; and it appears from the Report of the British Association of Science in 1847, that in the ninth and tenth centuries the Danes had fisheries on the coast of that island, from whence they were accustomed to send large exports to the south of Europe. In the reign of the English Queen Mary, Philip the Second of Spain paid the sum of one thousand pounds yearly for securing to the Spaniards the right of fishing on the Irish coast; and the Dutch purchased a similar right from Charles the Second at the price of thirty thousand pounds. It was also granted as a favour to the kingdom of Sweden, in 1650, to employ one hundred vessels in the same pursuit; but long before this, in the reign of King John, the merchants of Bayonne, who already rented from the English crown the right of taking Whales in our seas, paid to the king six marks for the sole right of the trade of drying Congers and Merluciones on the English shores. Lysons supposes that these Merluciones were Whitings and Haddocks; but the Merlucius, Sea Pike, or Sea Luce of ancient authors, is represented without a barb at the lower jaw, and with only two fins on the back; which circumstance, coupled with our knowledge of the great fishery carried on expressly for Hakes, is sufficient to determine the species. Under the name of Merluce, or Sea Pike, this fish also occupied a station in heraldry.

It may be here incidentally remarked also, as an illustration of the violent stretches of royal prerogative, by which the industry of their native subjects was cramped for the benefit of strangers, that a fishery for Whales on the English coasts was the subject of a similar grant; by which King John assigned over to the merchants of Bayonne, at the price of ten pounds yearly, the exclusive right of taking these creatures in all the space between St. Michael's Mount and Dartmouth. In the last-mentioned case it is probable indeed that the practical hardship was not great, and the plea may have been urged that the objects of pursuit were by prescription a royal possession; for the reason has been assigned, that the king ought to claim the head of a Whale, that his lamps might be supplied with oil, whilst the bones of the tail, (if such could be found!) were claimed by the lawyers as necessary to the symmetry of the queen's dress: a precursor, it seems, of the crinoline of our own day.

Whether we are to ascribe it to the superior quality of the fish, or to a difference of taste in the people, there are even now places where the Hake is regarded as among the most valued of fishes, and where in consequence it bears a high price; for whilst with us the Hake of perhaps a dozen pounds in weight, will be sold for sixpence or less, we have heard of half as many shillings as paid for it in Portugal. It seems also to be held in estimation at the Cape of Good Hope, where, according to Dr. Pappe, this fish was not known before the occurrence of an earthquake in December, 1809; and where at first it was so scarce as to be sold at the price of four shillings and sixpence. Since that time it has increased in that neighbourhood yearly, and at this time is caught in such abundance as to afford a considerable quantity for exportation. In this case, however, a question arises with regard to the identity of the species with our own; but it is to be remarked that Lacepede informs us he had discovered among the manuscripts of the well-known naturalist Commerson, the mention of a fish which bore all the characters of our Hake, and which he had met with in the waters of the Southern Ocean. I have been informed also that it has been found on the coast of California.

Hakes when salted and dried, without the head and a portion of the back-bone, are among the stores laid up by fishermen for the sustenance of their families when prevented from following their usual employment in stormy weather; or when in the early months of spring the larger number of our fishes have gone into deep water beyond the reach of their lines.

The length of a Hake may come near to four feet, and a large one has been known to weigh twenty-two pounds. Body and head moderately lengthened, thick; head wide between the eyes, moderately compressed at the sides; snout projecting, broad, bony; under jaw longest. Gape wide; teeth strong and prominent, arranged like a horse-shoe in front of the palate. Eyes moderate; nostrils nearer the eye than the snout. The posterior plate of the gill-cover oval, behind the root of the pectoral fin; the divisions of the gill-covers well marked. The lateral line runs high at first, and in a young example passes on to the caudal fin obliquely downward. Scales on the cheeks and body, larger than in most of this family, and firmly fixed. The first dorsal fin begins a little behind the root of the pectoral; second dorsal and anal more expanded towards their termination, their rays more plainly marked than in others of the family, the connecting membrane thin; and although in full-grown examples these rays are divided into branches towards their points, it is not so in early growth, for at this stage they are firm and simple, about half grown when they become bifid, beginning with those nearest the tail; tail and pectorals slightly round; of the ventral fins the fifth and sixth are the longest. Colour generally dull brown on the back, but when in the best condition rich brown with a tint of purple; sides lighter, sometimes pink, dull white, or bright below; yellowish on the cheeks; lateral line yellow or brown. Dorsal and pectoral fins and tail dark; anal fin whitish near the vent, darker behind. The first dorsal fin has ten rays, second thirty-eight, pectoral thirteen, anal thirty-eight, ventral eight, tail twenty-two.

We have shewn, when treating of the Haddock, that the fish called Onos and Asinus by the Greeks and Romans was the Hake, and not the former species, as many have supposed. It remains to be seen whether Swainson's opinion is correct, that the Hake of the Mediterranean is a different fish from our own.

MOTELLA.

The definition of this genus by Cuvier is,—the anterior dorsal fin so low as to be scarcely perceptible, which, standing alone, is at least as applicable to the very different genus *Raniceps* as to this. But the organ often represented as the first dorsal fin, in this genus has no nearer title to the name of a fin than what arises from its situation at the anterior portion of the back, where its seat is in a chink, from whence it projects when the fish is in the water, but it lies almost hidden when the fish is dead. It does not possess rays, as do all the true fins of these fishes; but it is formed of membrane, from the edge of which rises a thickly placed row of threads, the foremost of which is the stoutest and most prominent. When these fishes are alive in their native element, and resting, as they familiarly do, on the ground, with all their true fins at rest, this organ is in continued and rapid action; and its intimate structure shews that while it is destitute of any power of propulsion, or of regulating motion, it is well furnished with nerves which render it acutely sensible to impressions. This dorsal membrane is in fact supplied with a special nerve, which reaches it directly from the brain, and which also passes onward to the true dorsal fin. A branch of this nerve also goes to the pectoral and ventral fins, which are thus endued with particular powers of sensation, in addition to those of action, the last-named faculty being influenced into energy by branches of the intercostal nerves. The facial nerve also divides into four branches, one of which passes to the back of the head and upper portion of the outside of the orbit; a second branch passes through the lower portion of the orbit, and also supplies the director muscles of the eye itself, whilst the other two are distributed to what may be spoken of as the face. These observations have been obtained from examination of the Three-bearded Rockling, *Motella vulgaris* or *tricirrata*, but it is probable that they are equally applicable to the other species of this family; and they tend to shew the exquisite provision which is made for the supply of organic sensibility to a family of small fishes, the lives of which are for the most part passed in places where feeling is more important even than sight, as well for their safety as subsistence. The essential character of the genus *Motella* is, that with the other portions of structure of the gadoid family, they possess barbs on the upper and lower jaws, together with a ciliated membrane, which is seated in a chink between the head and dorsal fin.

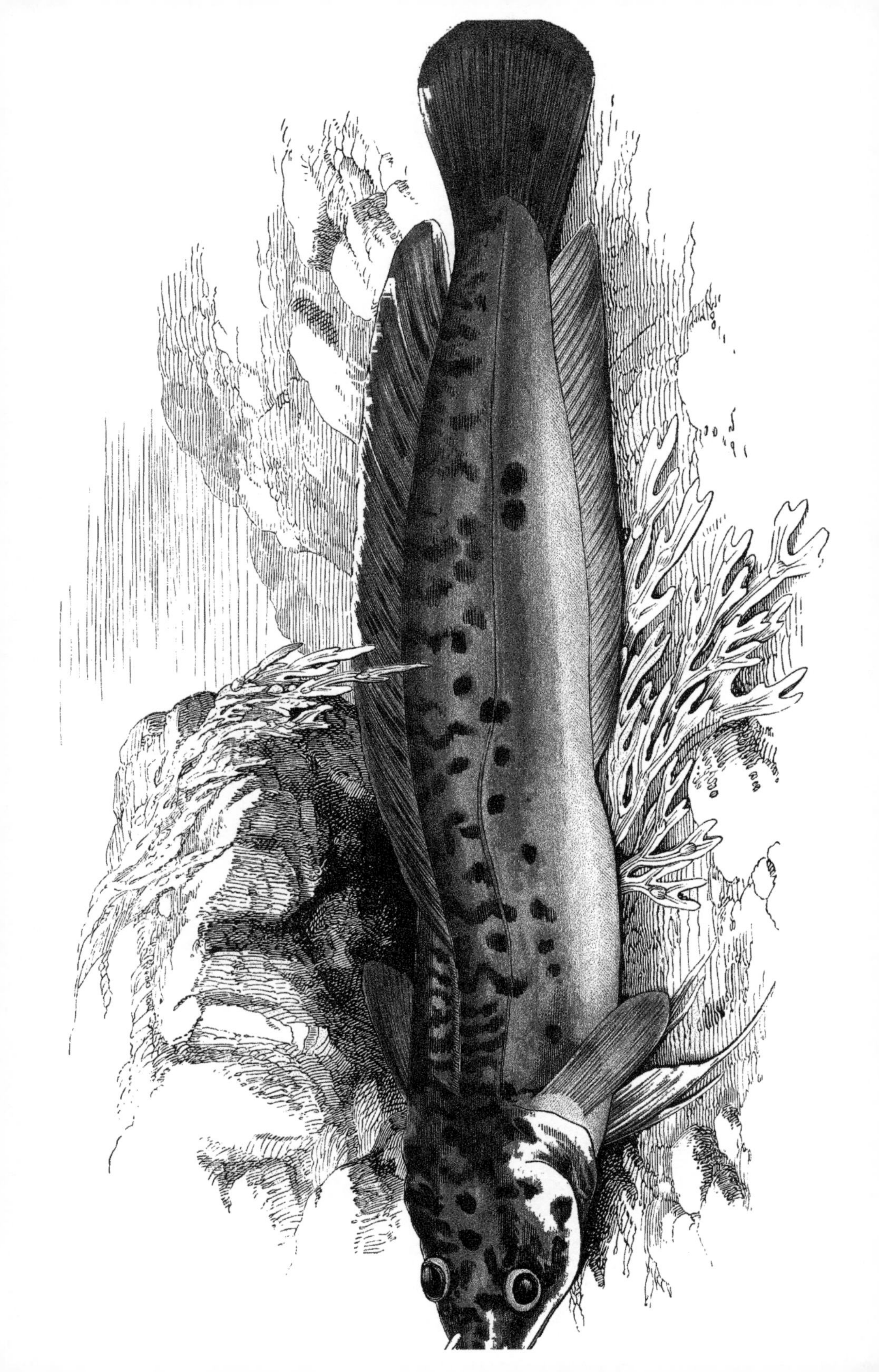

THREE-BEARDED ROCKLING.

WHISTLER. WHISTLE-FISH.

Mustela vulgaris, Whistle-fish,	WILLOUGHBY; p. 121, Table H. 2.
Gade mustelle,	LACEPEDE. RISSO. BLOCH; pl. 165.
Gadus tricirratus,	DONOVAN; pl. 2.
" "	FLEMING; British Animals. p. 193.
Motella tricirrata,	JENYNS; Manual, p. 449.
" *vulgaris,*	YARRELL; Br. Fishes, vol. ii, p. 270.
" "	GUNTHER; Cat. Br. M., vol. iv, p. 365.

UNTIL of late writers on natural history have regarded this fish and that which possesses four barbs on its snout, and consequently five in all, as only varieties of the same species; and they may stand excused for so doing, in the consideration that the examples of each, when of smaller size, and keeping closer to the lower tide-mark, as the Five-bearded Rockling generally does, are found to resemble each other closely, except in the particular that the last-named is furnished with a different number of barbs. But recent observation shews that these two fishes are naturally distinct, as well in their habits and distribution as in the obvious particular of a variety in the number of the processes or barbs.

The Three-bearded Rockling is often found where sea-weeds cover oozy ground, and there it hides itself under the shelter of a stone when the tide has retired. Under these circumstances the specimens are distinguished by uniformity of colour; the back and sides being chesnut brown, which is softened into yellow on the under parts. But those examples which are met with at perhaps the depth of ten or twenty fathoms are of much larger size, and their colour shews considerable difference; the ground of it being pale or reddish yellow, studded with spots of deep brown of the same tint as that which covers the body of the smaller individuals near the shore. And that they go to even a greater depth than that we have mentioned appears from the circumstance of an example which was found in the stomach of a fish caught at the depth of forty fathoms; but the rarity of such an instance tends to shew that so great a distance from land is not their usual resort. It is when they have reached their largest size that they become of importance to fishermen on some parts of the coast, where they are valued for the table by gentlemen who have learnt to esteem them as a delicacy. Their station is always at the bottom, where their food for the most part is the smaller crustaceans and worms; but in the month of January, after stormy weather, there were found in some that had been thrown on shore, entangled in loose sea-weed, not only crustacean animals, but loose pieces of brown sea-weed.

This species is rather widely distributed, so that it is found as far north as Sweden, and it is set down by Mr. Lowe amongst the fishes of Madeira. It is also common in the Mediterranean, where it has been judged to form one of the fishes named by the ancients Asellus, (Pliny, B. 9, C. 23.) Cuvier says that nearly all naturalists, after the example of Rondeletius, have applied this name (of Asellus) to the Merlus, the *Gadus merlucius* of Linnæus, or Hake; and he appears to take credit to himself for saying that he has found only one fish to which he might apply the character of the ancient Asellus; and that is the present species, *Motella tricirrata*, or Three-bearded Rockling. But on the other hand we find Jonston anticipating Cuvier, in quoting Rondeletius for the same opinion; and yet on referring to Rondeletius, we find him limiting his remark by saying that this Rockling is the lesser Asellus, or *Callarias minor*, and that his other kind of Asellus, or *Callarias bacchus*, is the *merlucius*, or Hake. Why this last kind of Asellus was called *bacchus* we leave to conjecture; but it may have been that as it was mostly used in a salted state by ordinary persons. it produced the effect of sending them often to the wine shop.

This fish spawns about the end of winter; and I have known it large with spawn about the end of April.

The Three-bearded Rockling in its general shape resembles the Ling, and sometimes attains the length of fourteen inches; head flattened on the top, a little compressed at the sides; under jaw shortest, teeth in both, a triangular arrangement of them in the palate. Lower jaw with one barb, the upper jaw with two, with the nasal aperture at their base. These barbs, always those in the five-barbed species, when the fish is alive, as also project straight forward; and when, as in many figures of these fishes, they are represented as limp or crooked, it becomes certain that the drawing was made from a dead, and perhaps from a stuffed example.

The eyes are large and prominent. Dorsal and anal fins long, both of them ending near the tail, with a small degree of expansion at that part. A deep chink behind the head, which is the seat of a membrane that has been described as a first dorsal fin, furnished along its edge with numerous slender filaments, and a separate stouter one is placed before it. This form of membrane, characteristic of the genus, is accompanied with a structure enclosed within the substance of the flesh, which goes still further the difference between it and the ordinary nature of a fin. In every case of a dorsal fin there is between the upright spinous processes of the vertebræ and the true rays of the fins a row of intermediate bones, on which the latter rest, or to which they are attached, and by means of which, with corresponding muscles, their motions are regulated. But this vibrating membrane has no connection with such a row of bones, but in the place of them there are very slender perpendicular muscles, the ends of which are attached to the processes of the vertebræ, and at the other extremity are lost in the substance of this membrane. The use of these muscular fibres is obvious. Vent a little before the middle of the body. The lateral line is raised at first, and sinks a little in its progress. Pectoral fins round; ventrals lengthened and pointed, in some larger examples the two first rays separated for a long space, and the second ray the longest. The colour of the smaller and in-shore examples has been already specified. Of those of larger size few are spotted alike, and in some on the anterior portion of the back the deeper colour bears the appearance of bands. The barbs and borders of all the fins are often of a bright red.

It is to be observed that Dr. Gunther represents the Rockling marked with "a row of brown spots along the base of the dorsal fin," and, as he says, with teeth a little different, as a separate species, under the name of *Motella maculata*.

FIVE-BEARDED ROCKLING.

Five-bearded Rockling,	WILLOUGHBY; p, 121.
Gadus mustela,	LINNÆUS. FLEMING; Br. Animals, p. 193.
" "	DONOVAN; pl. 14.
Motella mustela,	JENYNS; Manual, p. 450.
" *quinquecirrata,*	YARRELL; Br. Fishes, vol. ii, p. 278.
" "	GUNTHER; Cat. Br. Museum, vol. iv. p. 364.

THIS species bears so near a resemblance to the fish last described, as well in shape as colour, at least in the earlier stages of growth of the latter, as also in its general habits, that it appears unnecessary to enter on a detail of anything concerning it more than the particulars in which they are found to differ; and the only exception we have to make to this is in the account we shall give of the way in which the spawn is deposited, of which it remains a matter of doubt whether what has been observed belongs to one or the other of them, or whether both of them pursue the same habit.

On our shores in the earlier condition of the Three-bearded Rockling, and until it reaches to about the length of six or eight inches, they are found in the same places, in like abundance, and bearing the same appearance; but I have never obtained an example from such a depth of water as the larger individuals of the three-bearded species, or, as Dr. Gunther describes it, the Spotted Rockling, *M. maculata,* is known to inhabit; nor has it happened that larger specimens have been met with than of the length given above. It is enumerated among the fishes of Scandinavia by Nilsson, and as existing in the Mediterranean by Risso, who indeed regards it as only a slight variation of the kindred species; but it is not mentioned among the fishes of Madeira by Mr. Lowe.

The following observations as regards the care which this fish bestows on the safety of its spawn, were made by my now

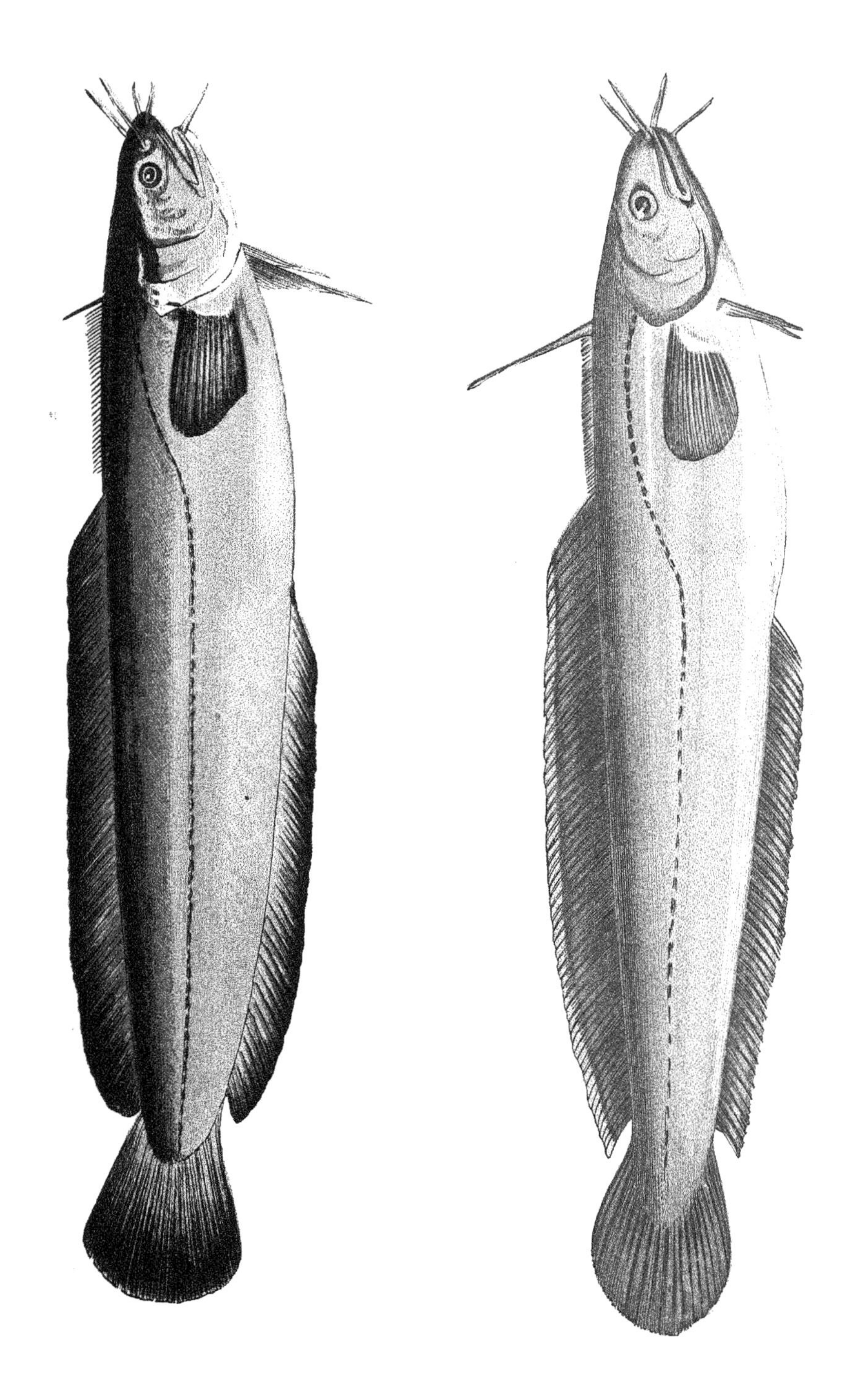

deceased son Mr. Richard Quiller Couch, at the time when he lived at Polperro; and when he was engaged in watching a similar proceeding in the Fifteen-spined Stickleback. He says, —The next nest shews considerably less skill, but more perseverance and energy. It is invariably formed of the common coralline, (*Corallina officinalis,*) in large quantity, put together without skill or arrangement, thrust into some cavity or crevice of a rock close to the low-water mark, and the materials are maintained there by no other bond than that of compression. And as the coralline of which it is formed is sometimes not to be found within the distance of one or two hundred feet of the selected spot, it must be gradually collected, and brought with a degree of perseverance at least equal to the intelligence displayed in the construction. But perhaps the most extraordinary part of the proceeding is shewn in the force exerted by the fish when thrusting it so firmly into the crevice or hole in the rock, and which we should have judged to be beyond the power of any fish we are acquainted with. The grains of spawn are small, their size being about the fifteenth of an inch in diameter; semitransparent and yellow. They are not contained in a cavity, like those of the Stickleback, but are scattered through the mass, sometimes indeed in clumps, but at others irregularly on the coralline.

From the compact character of the nest, and that the grains are dispersed through all parts of its structure, it is evident that these grains of spawn must be deposited while the nest is in the progress of formation. Having preserved the ova in water until the young had come to life and escaped from their confinement, in order to be assured of the species to which they belonged, the conclusion drawn from their shape and spotted appearance was that they were the progeny of a species of Rockling; but on this point a less amount of certainty was felt, as from the almost inaccessible situation of the places in which the nests were placed, they could not be strictly watched when the tide had flowed sufficiently high to cover them. Whether any of these nests were permanently covered with the sea on the rocky coast where they occur remains uncertain, but with reference to the doubt here expressed with regard to the species of fish produced from these ova, on close observation I felt no doubt that they were of one or other of these species of Rocklings, although the parents themselves were not seen in attendance. These observations were made in the spring; but it must happen that these nests are formed at different times, or that the young ones come to life in long succession, for at the middle of July some of them are barely an inch in length, while others are at least three times as long. Both these Rocklings will take a bait.

With regard to description this species is readily distinguished by the presence of four barbs in front on the upper jaw, and one below. Two of these barbs above proceed from the borders of the nostrils; but in some examples from the north of Somersetshire, with which I was favoured by E. T. Higgins, Esq., a low membrane proceeded from each pair of the upper barbs, and united them together, which is not the case with these fishes as found on the south coast; yet I cannot on this account as yet regard them as separate species. In one instance which fell under observation the barbs on one side were united into one, while on the other side they remained separate. This fish is too small to be regarded as an article of food, but it might be employed by way of bait for other fish if the numbers were more abundant, or the fishermen could be persuaded to use them.

FOUR-BEARDED ROCKLING.

Gadus cimbrius,	TURTON'S Linnæus.
Gade cimbre,	LACEPEDE.
Motella cimbria,	YARRELL; Br. F., vol. ii, p. 274.
" "	GUNTHER; Cat. Br. Museum, vol. iv, p. 367.

THIS is particularly a northern species, but although met with on the coasts of Norway and Sweden it remained unknown to Linnæus, at least to so late a date as the publication of the tenth edition of his System. It was first recognised as British by Dr. Parnell; but although in some places it is not uncommon, its distribution, even in the north of our islands, appears to be limited to certain districts; which circumstance may be caused chiefly by a congenial nature in the ground; but the reason of its remaining little known in some fishing stations while it is common in others, may arise from a difference in the method of fishing; for as its favoured residence is in waters of considerable depth, it can be taken only where long lines or bulteys are employed, the hooks of which, although generally large, it is able to swallow. On inquiry it has been found that there are stations in Scotland where this fish is unknown; but on the other hand, it is set down by Mr. Cocks as among the rarer fishes which he has seen so far to the south and west as Falmouth. In its proper haunts it keeps near the ground, where it feeds on such crustaceous animals and worms as are there common; but it is probable that like the others of its family, its appetite is exercised without much discrimination.

A close description of the Four-bearded Rockling seems unnecessary, as in shape it bears a near resemblance to the three and five-bearded species, while it is the only one of this family we are acquainted with that is furnished with three barbs on the upper lip, and one of them as it stands alone projects exactly in front. In this respect its structure may be easily distinguished from that irregular formation which has been referred to, and which may not occur again in the Five-bearded Rockling. A further distinction is noticed in the much longer extension of the thread, or process, which stands up before the ciliated membrane behind the head. The ventral fins are shorter than in the other species, and the head somewhat longer. It is probable that its colours vary in some degree, according to the nature of the ground; but usually, while the back and sides are dusky and reddish, there is a whitish border to the dark brown dorsal and anal fins. It has been known to measure fifteen inches in length.

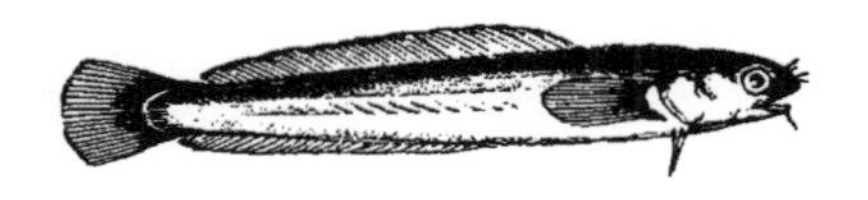

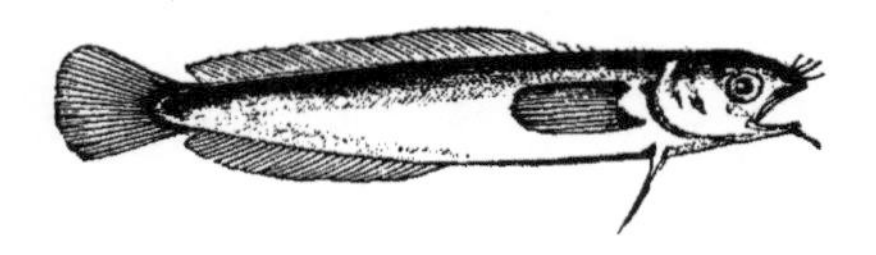

1 & 2 MACKAREL MIDGE.

3 & 4. THOMPSON'S MIDGE (3 Natural size.)

CLI

COUCHIA.

When a notice of the most common species of the fishes comprised in this genus was published in the fifth volume of Loudon's "Magazine of Natural History," the name assigned to it was *Ciliata*, as being expressive of the membranous organ bordered with slender threads in front of the dorsal fin, which is common to this new genus, with that of *Motella*; but as this name is found to have been already appropriated to a different object, it pleased Mr. Thompson, of Belfast, to substitute for it the term *Couchia*, which has received an establishment by being adopted by Dr. Gunther in his "Catalogue of the Fishes of the British Museum.

The definition of this genus is,—The body compressed, moderately elongated, covered with minute scales, and with a silvery pigment inferiorly. Head compressed, with the upper jaw longest. A separate caudal, anterior and dorsal membrane, which is composed of a band of numerous short fringes, concealed in a longitudinal groove, the first fringe prolonged. One anal fin; ventrals composed of about four rays Minute teeth in the jaws and on the vomer. Snout with barbles. Air bladder none. Pyloric appendages in small number. It will be observed that in his definition Dr. Gunther regards the fringed membrane on the back as a fin, which appears to be more than doubtful; but the distinction of the genus is well made out, so as not to be confounded with any other.

MACKAREL MIDGE.

Ciliata glauca,	Couch; in Loudon's Magazine Nat. Hist., vol. v.
Motella glauca,	Jenyns; Manual, p. 451.
" "	Yarrell; Br. Fishes, vol. ii, p. 281.
Couchia minor,	Thompson; Nat. Hist. Ireland, vol. iv, p. 190.
" *glauca,*	Gunther; Catalogue Br. Museum, vol. iv., p. 363.

This is one of the smallest fishes of which I have any knowledge, so that even compared with the usual magnitude of the Gobies, it hides its diminished bulk; but on the other hand, when brought into comparison with the mighty whale—that sea beast

———"which God of all his works
Created hugest—that swim the ocean stream,"

the beauty and complexity of its structure serve well to illustrate the creative energy of that BEING, who is as great in the small as in the most conspicuous, and who renders all things perfect that proceed from His hand. Yet not such appears to have been the opinion of some naturalists, who, from the minute size of this species, could not persuade themselves that it was other than some early stage of existence of some unknown larger fish. It was again a similar cast of thought that led the older naturalists to believe that apparently these and some others of small size, which appeared at times in vast numbers, were produced by spontaneous generation from the froth of the sea or the putrefaction of marine substances, and not from eggs like the larger, and as they judged, more elaborate kinds; on which account they bestowed on them the name of *Apua*.

But however small, this species is far from being the least useful of the natives of the ocean; for, as the prey of other fishes, it helps to form a powerful attraction which draws many of the wandering and more valuable kinds within the reach of human effort, and by so doing afford employment and subsistence to large numbers of people, who know nothing of that economy of providence by which their wants are abundantly supplied. The Mackarel Midge is widely dispersed over the Atlantic Ocean, for it is recognised by Nilsson among the fishes of the Northern Sea, and through the kindness of Mr. Higgins I have received examples from the bay of Chesapeake in America. On our own coasts they perform a sort of migration, or rather change of quarters; which appears to be chiefly from the deeper to the shallow water, and from the bottom to the surface. They are not recorded as presenting themselves to the notice of fishermen generally until about the middle of May; after which time they often abound at all distances from land to which fishermen are accustomed to proceed; and it is only late in autumn that they cease to be seen. Through the warmer seasons of the year they keep close to the surface in small companies, and in stormy weather it is common for them to be thrown on board a fisherman's boat from the crest of a wave when it breaks over the gunwale, or perhaps on the beach entangled in sea-weed. As if conscious of danger from every prowling inhabitant of the deep, these fishes are ever ready to seek the shelter of a floating object; and they appear to feel pleasure in rubbing themselves by passing to and fro below a floating clump of sea-weed, or the corks of a net. They also rejoice in the shelter afforded beneath the expanded canopy of the larger medusæ; and in keeping close to such objects they sometimes become so entirely off their guard as to suffer themselves to be taken with the hand. Yet they usually exercise much vigilance and activity, and dart away on a slight appearance of danger; and when left by the tide in some large pool of the rocks, their movements are nimble, with the appearance of being instigated by intense feelings of delight.

Pursuing and pursued the tail has been bent to one side, and in this position it has urged and driven the little creatures in their gambols, safe, at that time at least, from the intrusion of voracious enemies. Kept in a glass vessel, there is shewn the same disposition to seek the shelter of some covering object, whether on the surface or at the bottom; and in the situation last mentioned, the fish was so fortunate as to find a small fragment of green oreweed, beneath which it passed a considerable portion of its time.

They die immediately on being taken from the water. There is no doubt that these fishes feed on animal substances, and there is proof that they are even eager after it. An individual of this species was placed in a glass globe with two very small Grey Mullets, and a piece of brown oreweed that was covered with cord-like convolutions of the ova of some kind of molluscous creature of a pink red colour. After a fortnight it was found that these mullets had devoured the whole of the ova contained in this mucous cord, but the covering itself was only torn to pieces by them; but the Mackarel Midge had had no share in this feast. His longing was for more substantial things, and he finally attacked one of the Mullets, which he laid hold of near its head, and so grasped it as to carry it about his prison for more than a minute, without being able to swallow the fish, which was equal to two thirds of its own size. Both it and the surviving Mullet were afterwards fed on bread. Mr. Peach discovered this fish in abundance at Wick, in Scotland; and he found that they took animal food freely in captivity, making a dash at it when held to them on a feather; and so firmly did they hold it, that they were lifted out of the water as they retained it with their teeth, fighting with each other for the prize, and shaking their heads to prepare it for being swallowed. Like the Rocklings when in the water the ciliated membrane is kept in constant and rapid motion, and it is only when thus situated that the position of the barbs can be well observed.

This fish must be very prolific, if we may judge by the multitudes which sometimes appear, and of which we have reason to believe that vast numbers fall a prey to the more ravenous fishes. They have even been found in the stomach of the apparently harmless Pilchard; but the time of shedding the roe has not been noticed. When they shew themselves with us they appear to be of full growth, and they rarely exceed the length of an inch and one fourth, the general proportion of the body, as compared with the Rocklings, being like that of the Whiting, excluding the fins, when laid by the side of the Common Ling. The head obtuse, compressed, upper jaw longest, with four projecting barbles; under jaw with one barble; teeth in both, and in the palate. Eye large and bright; behind the head a chink holding a fringed membrane. Dorsal and anal fins single, reaching to near the tail; pectoral and ventral fins rather large for the size of the fish; scales easily rubbed off. Colour on the back bluish green, sometimes blackish; belly and fins brilliant white or silvery.

It scarcely appears necessary to give a separate notice of a fish which was first described by Montagu, and termed by him *Gadus argenteolus*—the Silvery Gade, but which he seems to have confounded with the species we have last described, since he represents it as occasionally common on the western coast of England, where, since the distinction has been made, it has been again recognised. The important mark of difference between the Mackarel Midge and Montagu's Silvery Gade is, that the latter possesses only two barbs in the upper jaw; but it appears also to attain a larger size, since Montagu's specimens were two inches in length, and Dr. Gunther mentions examples obtained from Greenland which measured three inches. The proportions of the head and body are described as the same in both, but the number of rays in the fins are said to be different; those of *Couchia glauca* being in the dorsal forty-four, anal thirty-eight, ventral three; but in Montagu's Silvery Gade—*Couchia argentata* of Dr. Gunther's Catalogue of the British

Museum, vol. iv,—the dorsal fin has fifty-eight, anal forty-four. In the ventrals there is the same number in both. Nilsson does not acknowledge these fishes to be distinct.

THOMPSON'S MIDGE.

Coryphæna.

It must have been this fish which Mr. Thompson has described in the fourth volume of his "Natural History of Ireland," and which he supposed to be our well-known Mackarel Midge, but on comparison with which, when authentic specimens of the last-named species were supplied to him, he felt himself in a state of uncertainty. We copy his lengthened account, as it supplies us with information regarding some of its habits and motions, which differ decidedly from those which have been noticed in the true Mackarel Midge, as they frisk on the surface of the ocean.

"Descriptions of a minute fish allied to the *Ciliata glauca*, Couch, and *Gadus argenteolus*, Montagu. Plate 16, f. 1, 2, 3, of Annals of Nat. History, vol. ii.

"When dredging in Strangford Lough, County Down, on the 2nd. of July last, at from one to three quarters of a mile off the shore, and the water from ten to twenty fathoms in depth, I for upwards of an hour remarked some very minute fishes coming singly to the surface. They ascended in a somewhat vertical direction, remained but momentarily there, and again, generally in a similar manner, descended until lost to view. Their back appeared to be of a dark colour, but their sides presented the brilliancy of the brightest silver. Their size was rather under an inch; their motion, though somewhat wriggling, surprisingly rapid, so much so, that although the boat was scarcely moving, and the sea quite calm, their continuance at the surface was so short, that the greatest activity had to be exerted to secure them. For this purpose a small canvas net, otherwise used in the capture of minute medusæ, was available. When brought into the boat. they at first sight called to mind the *Ciliata glauca* and *Gadus argenteolus;* but the great size of the ventral fins, which were likewise of a pitchy blackness for nearly the last third of their length, seemed opposed to their identity with these species. The boatmen who accompanied me had not observed this fish before, nor had they heard anything of it.

"The general form elongate; belly protuberant. On a close examination of all the specimens, nine in number, no cirri can, with a high power of lens or on the field of the microscope, be detected on either jaw. The largest individual, ten and a half lines in length, may be characterized as having the upper jaw the longer; strong and pointed teeth in both jaws; head occupying rather more than one fourth of the entire length; eye equal in diameter to one third the length of the head; opercle rounded at the base, altogether forming a portion of a circle; first dorsal fin originating just over the opercle, so sunken, and its rays (which are thick and blunt) so short, as to be hardly distinguishable in the profile of the fish, not less than twenty-five rays; second dorsal commencing close to the first, and before the end of the pectorals, of unequal height, extending to the base of the caudal, not less than fifty rays; pectoral fins rather less than one fifth of the entire length, of moderate size and rounded; placed very high, somewhat above the opercle, about twenty rays; ventrals placed high, commencing rather in advance of the pectorals, somewhat square at the end, occupying one fourth of the entire length, reaching to the vent and consisting of about six rays; anal fin commencing at the vent and extending to the base of the caudal, unequal in height, having at least forty rays; caudal fin elongate, occupying one fifth of the entire length—measured from the last vertebra of the body, somewhat rounded at the end, containing about thirty rays; branchial rays about seven; vent midway between snout and base of the caudal fin. Colour, (when recent) back rich green, varied with dots of gold and black; operculum, entire sides, and under surface bright silver; pectoral, dorsal, anal, and caudal fins uniformly of a pale colour; ventrals likewise so for two thirds from the base, remainder pitch black; irides silvery.

"Since the above was written I have been favoured by Mr. Yarrell with original specimens of *Ciliata glauca,* obtained from Mr. Couch, and from these the Strangford species differs as follows:—My specimens, under eleven lines in length, do not, like the Cornish fish, which is one inch five lines long, exhibit cirri on either jaw. The ventral fins in mine are equal to one fourth of the entire length, in the English specimens to about one seventh; in the latter the longest rays have a fibrous termination, whereas those fins are somewhat square at the end in the Strangford specimens; besides, they are in those of a pitchy blackness for the last third of their length, although in the other of a uniform pale colour throughout. These differences were likewise constant in Cornwall and Strangford specimens of similar length."

This minute description by Mr. Thompson will render unnecessary any further remarks of the same sort derived from specimens which I have obtained, and for which I return thanks to Mr. Thomas Edwards, of Banff, in North Britain; who in the capture of a single specimen readily discovered that it belonged to a species not generally recognised, and who, on my enquiring after further particulars, readily procured for my use eleven other examples; some considerable advantage to science arising from this supply being that it not only satisfied me on a close scrutiny that there did not exist any barbs where Thompson could not discern them, on the upper and under lips, but also that the ventral fins were not situated near the throat as that gentleman supposed, but that they were placed far back on the belly; and also not in near contact with each other. On the point of situation repeated examination has compelled me indeed to form a different conclusion from that arrived at by Mr. Thompson; but I feel persuaded that this difference is to be regarded on his part only as a slip of the pen, and the whole of the other particulars are too much alike to admit of a doubt that the Scottish fish and that of Ireland are exactly the same. Laid by the side of the true Mackarel Midge, the difference between them is conspicuous; and the greater length of the ventral fins, with their situation and intensely black colour of the hindmost portion were obvious in all the specimens. No ciliated membrane could be discovered.

A later communication from Mr. Edwards confirms the remark of Mr. Thompson as regards the surprising agility of these fishes; their nimbleness being such as to render it exceedingly difficult to catch them with a net. He further remarks of an example which he kept alive for a few days that it displayed excessive restlessness and watchfulness; and when noticed in its native element on sandy ground with a flowing tide, although at times a wave might bear it further in, it presently made its way outward again to the distance of about a yard from the shore; and it was only by wading into the deeper water beyond it that it was at last secured. More were afterwards secured; but they disappeared suddenly, as if in periodical migration.

As it would scarcely have been safe to place this little fish in our British Catalogue as a distinct species, without first soliciting the opinion of Dr. John Edward Grey, of the British Museum—to whom we have had occasion to feel greatly obliged on other occasions—some examples were submitted to his inspection, and the following is a portion of his reply:—"It seems to be the young state of the genus *Coryphæna* or Dolphin. We have some specimens of twice the size of those that Mr. Edwards sent, and others intermediate in size between them and the adult fish. It is curious that the young *Coryphæna* should be found on the coast of Banff in abundance, and the adult not found there, as far as I know." The occurrence of this fish on the coast of Scotland is indeed remarkable, and especially as the observations of Mr. Thompson on the coast of Ireland tend to shew that the former is not an isolated instance. It is only provisionally that we have designated these examples by the name of Thompson's Midge.

RANICEPS.

The head large, broad, and depressed; body of moderate length, with very small scales. A barb at the lower jaw; small teeth in the jaws and middle of the palate, mingled with stronger ones in the former. Two dorsal fins, of which the first is exceedingly small; a single anal fin, both disjoined from the tail. Ventral fins jugular, with two of its rays much longer than the other.

LESSER FORKBEARD.

TADPOLE FISH.

Barbus minor, Lesser Forkbeard,	Jago, in Ray's Synopsis.
Blennius trifurcatus,	Turton's Linnæus.
Batrachoiide blennoide, and Blennie tridactyle,	Lacepede.
Raniceps trifurcatus, and R. jago,	Fleming; Br. Animals, p. 194.
" " "	Mag. of Zoology, by Sir W. Jardine, Bart., vol. i.
	Jenyns; Manual, p. 453.
"	Yarrell; British Fishes, vol. ii, p. 292.
Raniceps trifurcus,	Gunther; Cat. Br. Museum, vol. iv, p. 367.

This species was first made known to science by Jago, and was long believed to rank among our rarest fishes; but, since attention has been more generally directed toward the inhabitants of our seas, it is discovered to be not uncommon on all the coasts of the British islands, where the water is not shallow, and the bottom is rough or rocky.

Yet it appears to be a wandering as well as solitary fish; and although several have been taken in a season in the west of Cornwall, it is only on one occasion that I have

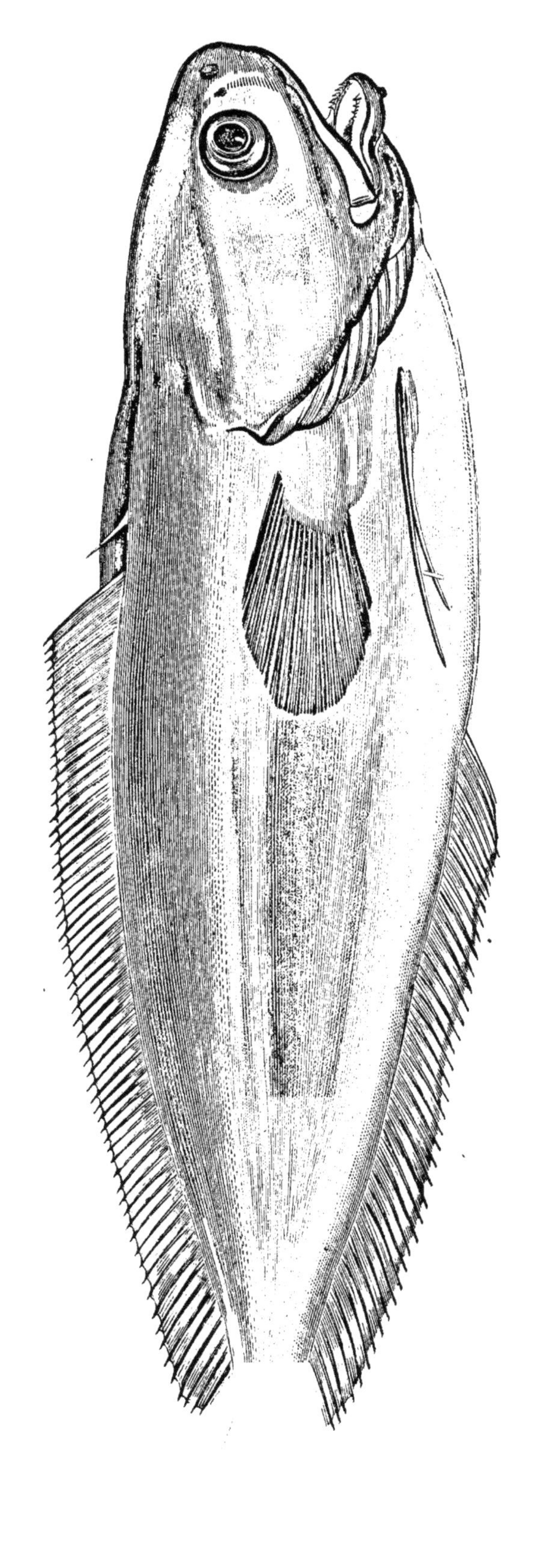

known a couple to be caught in one boat. It is probable that it feels some influence of the seasons in its change of quarters, and the impulse of spawning may also govern its actions, as is the case with so large a proportion of other fishes; but its coming nearer the land cannot be altogether for this latter purpose, since I have known it caught in the middle of August, early in September, in January, and also in April: in the last-named instance, when two or three examples were taken, the roe appeared to be at the beginning of its enlargement. That the young ones are produced at no great distance from the coast appears probable, although they are rarely met with; but Mr. Newman found several of small size among sprats in London, and I learn from Charles H. Gatty, Esq., F.L.S., that he obtained a small example at Great Yarmouth. Nilsson classes this fish among those of Scandinavia, and a representation of it, with some variation of colour from our own, is given by Fries and Ekstrom, in their beautiful work on the fishes of that country. This fish is not wanting in voracity; for, besides that most of the examples which have been caught, were taken with a hook, fragments of an *echinus* and the scales of a sprat have been found in the stomach. A fisherman informed me that when drawn up with a line it shewed itself to be a wild and active fish. A few hours from the time of being caught, the smell becomes strong and peculiar, not unlike that of the Rocklings.

This fish seldom exceeds the length of twelve inches: the head short, bulky, depressed, wide, rounded in front; eyes large, prominent, forward; nostrils still nearer the snout; under jaw shortest, with a short and thick barb. Teeth in both jaws and the roof of the mouth, sharp and incurved. A depression along the top of the head, separating the eyes, which are wide asunder. Body wide, more compressed, and becoming smaller towards the tail. First dorsal fin very small, pointed, opposite the root of the pectorals; second dorsal higher than the first; anal fin beginning opposite the termination of the pectoral, and, passing along parallel with the second dorsal, both end nearly together a little short of the tail. Pectoral fin round; ventrals with about six short rays, and two longer; in a fish measuring ten inches the longest of these rays being two inches, and both divided for about half their length; tail narrow, round; all the fins covered with the common skin. The colour varies, in some examples being nearly black above; in others dark brown, lighter on the belly; lips and mystache altogether white. Some doubts have existed in describing this fish, from the circumstance that in some instances it is said to have a row of prominences or tubercles above the pectoral fins, while in others the surface has been found entirely smooth; and on account of this variation of description, Dr. Fleming represents them as of two species, with the names of *R. trifurcatus*—the lateral line tuberculated, and *R. jago*—with the lateral line smooth. But from examination of several examples by different naturalists, no doubt remains that these supposed species are the same. In some instances these supposed tubercles have been visible when the fish was first caught; in others they have shewed themselves only after the lapse of a few hours, while in others they have never become visible or to be discerned by the touch. Fin rays, first dorsal three, second dorsal sixty-two, anal fifty-seven, pectoral twenty-two, ventral eight, caudal thirty-four.

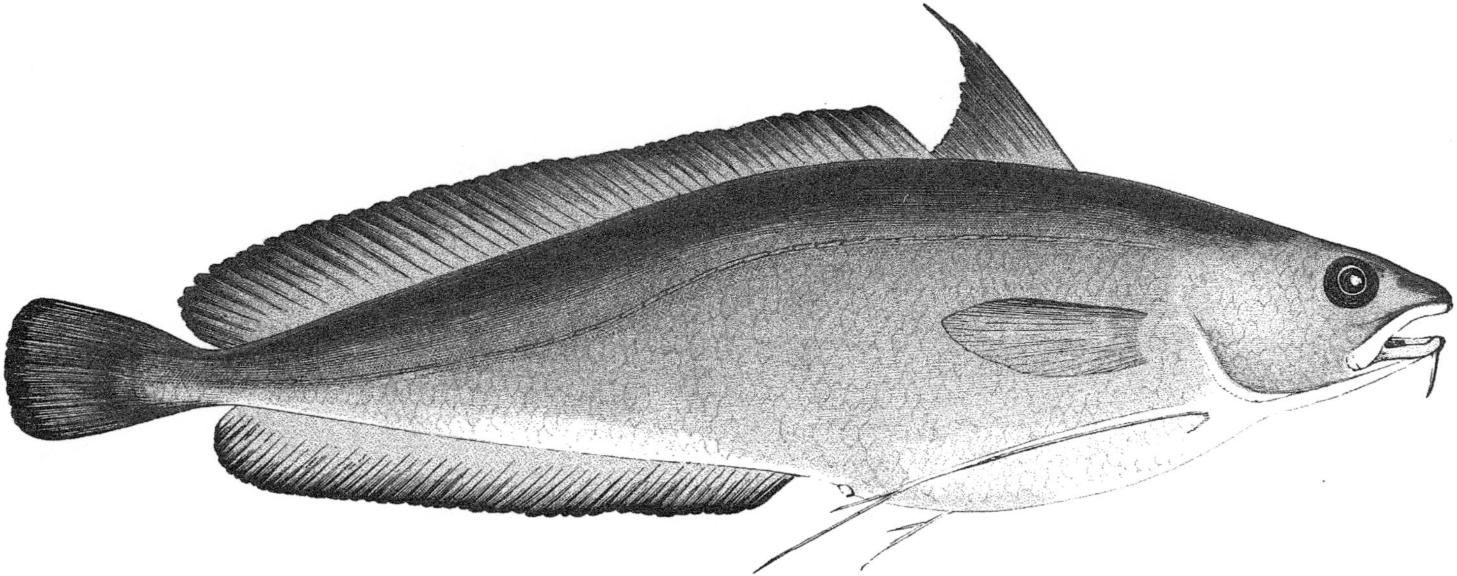

GREATER FORKBEARD.

PHYCIS.

Body of moderate length; jaws and palate with teeth; a barb at the lower jaw; two dorsal fins and one anal; but the distinguishing mark is—that the ventral fins are each formed of a single long ray which in the course of its length is divided into two.

GREATER FORKBEARD.

Hake's Dame, Forked Hake, Goat fish,	Jago; in Ray's Synopsis, p. 163 and f. 7.
"	Couch; in Transactions of Linnæan Society, vol. xiv.
Physcis furcatus,	Fleming; British Animals, p. 193.
Phycis furcatus,	Jenyns; Mânual, p. 452.
" "	Yarrell; British Fishes, vol. ii, p. 289.
Phycis blennoides,	Gunther; Catalogue British Museum, vol. iv, p. 351?

This species may be regarded as scarce rather than rare; so that examples show themselves singly, for the most part in the colder months, although I have obtained an example in June; and there are not usually more than one or two caught in a season. They are taken with a hook, and in the stomach I have found the bones of a small fish, a shrimp, and the fragment of a large pecten-shell. From the feeble structure of its tail it does not appear capable of active exertion, but this apparent defect is probably compensated by the structure of those long tendrils which stand in the place of the ventral fins which belong to the others of this natural family, and which in this instance may be judged from their structure to be endued with powers of lively sensation. These tendrils or fins have joints along their course, and are well supplied with nerves from what may be termed an axillary plexus, situated in the axilla of the fin, one branch of which passes along the course of the firm rays and sends off a branch to penetrate through it; while the other, which anastomoses with the first branch in the axilla, is carried along the posterior margin. These nerves are the largest I have found in the body of this fish, and their special function is shown by their proceeding from the spinal cord to their termination without communicating with any other nerve.

Considerable confusion and doubt have existed with regard to two or three species of this genus which bear a resemblance to each other; nor has the difficulty of distinguishing them been as yet cleared up. Rondeletius had long since given a figure of a fish of this genus, but as he was not acquainted with more than one species, and his plate is far from a good representation, there is little dependence to be placed on it as an authority on the question. But he gives a reason for the Latin and Greek name it bears, and which has now become the generic designation; and if we may depend upon the accuracy of the interpretation, which refers to the weeds of the sea, it will give us some information regarding its habits, of which otherwise little is known. He informs us that it was well known to fishermen that this Forkbeard was in the habit of forming a nest for the hatching of its young, among or of sea-weeds, and himself affirms that he was a witness of the truth of the fact. It is perhaps to this that the translator of Oppian refers when he speaks of fishes, that

> "They too, who like the mournful halcyons breed,
> And form a floating nest of slimy weed."
>
> B. 1.

But there is not a word of this in the original, where the Phycis is mentioned without any additional remark. The halcyons are excluded, and the sea-weed does not float, but lies at the bottom. And it is not improbable also that the observations of Rondeletius may be intended for another fish of the same genus; for it was Mr. Swainson's opinion that in the Mediterranean there were perhaps several which closely resembled each other, and Dr. Gunther mentions one of them under the name of *Phycis mediterraneus*. It may be made a question perhaps whether the fish we shall describe next in order is to be regarded as a separate species; but in any case it is certain that this latter also is a native of the Mediterranean Sea, and therefore the circumstance of forming a nest of sea-weeds may be true of it also; as well as its name of Mole which it bears in the south of France, and which is expressive of the pulpy nature of its flesh. In a note of my own I have compared it to that of a ling; and looking at Jonston's representation, Table 31, of a species of this genus which he calls *Gobius bottatriæ* of Salvianus, it may be asked whether it be not the long-lost Goby of Martial, so much valued at Venice.

A specimen slightly exceeded the length of twenty-five inches; the head flat on the top, compressed at the sides, and small in proportion to the body. Eyes large, situated forward towards the snout; nostrils in a depression before them; gape wide, under jaw shortest, teeth fine in both, those in the palate stouter. A barb at the lower jaw. Body compressed, deep before the vent, more slender near the tail; both body and cheeks clothed with scales. Belly protuberant. Lateral line bent slopingly down at about half its length. Two dorsal fins; the first elevated and pointed; the second and anal long, expanded posteriorly, bound down towards the tail. Ventral fins jugular, a simple cord, with two rays enclosed in one case, which divide at about two thirds of their length, reaching fully to the vent: in a fish of the length of two feet, the longest portion measuring eight inches, and the shortest five and a half. The tail round; all the rays of the fins soft. Colour of the back and sides dusky brown, more or less deep; belly whitish; fins dusky purple, except the ventrals.

BLENNOID FORKBEARD.

Lesser Hake,	PENNANT; pl. 32, edition of 1770.
Phycis blennoides	GUNTHER's Catalogue British Museum, vol. iv, p. 351.

IT is the opinion of Cuvier as well as of Dr. Gunther that this and the last-named are the same species; and with such an authority we leave the subject as it stands; but it is certain that the aspect of these fishes, as I am accustomed to see them, is not a little different, as will be discerned from the figures we have given, which are those of an example that measured in length twenty-eight inches, and of a young one of the length of four inches, which was drawn up in the shell of a living *Pinna ingens* from the depth of about forty fathoms. It is remarkable that this young example, which will be described in reference to the larger specimen, and which, although injured, has been sent to the British Museum, with another of nearly like size mentioned by Mr. Yarrell, were taken in the same manner, although with the lapse of several years between, are the only fishes I have ever known to be so caught, although the *Pinnæ* are often drawn up by the lines of fishermen. The proportions of the body in what we term the Blennoid Forkbeard are more slender than what we have noticed in the Hake's Dame; the depth of the body in the latter in front of the second dorsal fin, where it is deepest measuring one fourth of the length from the snout to the root of the tail, while in the Blennoid species it is equal to five portions and three fourths of the same length. In the first named fish also the sloping forward of the body begins behind the first dorsal fin, but in the Blennoid fish it scarcely begins to slope until over the eye; the under jaw also appeared decidedly shorter in proportion, and the lateral line less bent in its progress. It is a subject of regret that the scales were

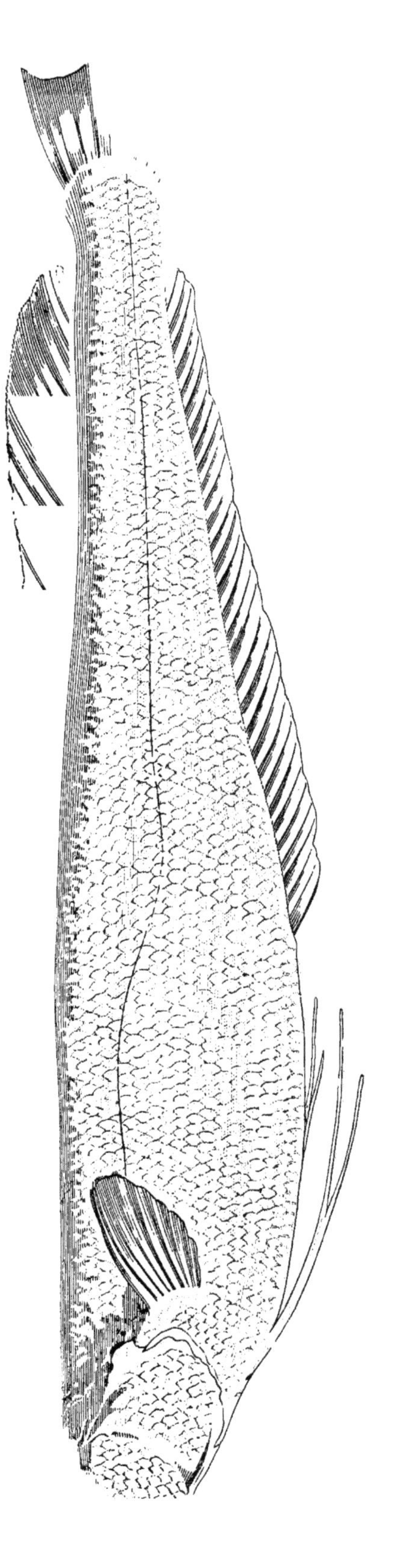

not in either instance counted, as Dr. Gunther lays much stress on their number in determining the species, and of which he observes that five or six series of scales are between the first dorsal fin and the lateral line. The ventral fins did not reach quite to the vent, and it may have been the mark of age that the barb at the point of the jaw was reduced to a stump; and while the tail was nearly even, a few of its upper rays were extended into a point. Visible scales covered the cheeks as well as the body, and the second dorsal and anal fins did not approach close to the tail. Colour generally light grey.

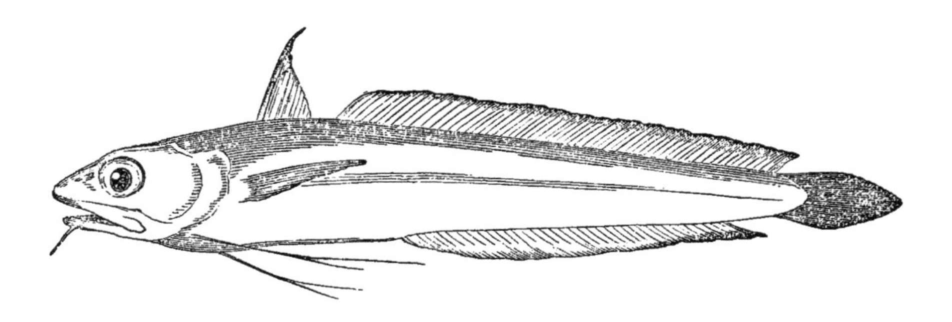

The young specimen, with the length of four inches, was not an inch in depth; snout much in front of the mouth, gape wide, opening back behind the front of the iris of the eye. Eye large, the two closer together than the breadth of either; scales rather large, vent a little nearer the head than to the root of the tail. Pectoral fin long and narrow, reaching to the origin of the second dorsal. First dorsal fin high, triangular, pointed, having ten rays, of which the first was single for a third of its length beyond the membrane. Second dorsal long, slender, pointed behind, the end of its rays reaching to the root of the tail; anal fin beginning close to the vent, and ending at some nearer distance from the tail than the dorsal. Caudal fin narrow and lancet-shaped. Ventrals a slender thread, reaching beyond the vent, and divided into two branches behind the half of its length. Colour of the fish light grey; anterior portion and end of the first dorsal black, its hindward border bright white, border of the second dorsal and tail black.

It has been already hinted that the Rocklings may have been the Asellus of ancient writers; but we shall only be doing justice to the subject when we remark further, that it is the opinion of Dr. Badham, in his amusing work of "Fish Tattle," that the kindred species *Phycis mediterraneus*—known by its lower and rounded first dorsal fin,—is truly the disputed fish; and if so, the present species will lay an equal claim to the designation, since both of them are natives of the same waters, and we are well aware of the little discrimination that was employed by the ancients in the distinction of species. It was their opinion that the tendrils of the Asellus were employed as bait or lines, and we have noticed in the Hake's Dame what organs of sensibility they are.

OPHIDIUM.

THE body lengthened, compressed; vent far behind. Gill openings wide. Dorsal and anal fins forming one with the tail: a bifurcated pair of tendrils hanging from the throat. Supposing these tendrils to be fins, Linnæus placed these fishes in his jugular order.

BEARDED OPHIDIUM.

Ophidion rondeletius,	WILLOUGHBY; p. 112, pl. 97.
Ophidium barbatum,	LINNÆUS. CUVIER.
Ophidie barbu,	LACEPEDE. RISSO.
Ophidium barbatum,	BLOCH; pl. 159. YARRELL; British Fishes, vol. ii, p. 415.
	GUNTHER; Catalogue British Museum, vol. iv, p. 377.

THIS fish is inserted in the British Catalogue on the remote and doubtful authority of former writers, none of whom profess to have seen a native example, or point out a place or time in which it has been obtained. Our figure, slightly tinted, is derived from Rondeletius, who knew the species well, as being frequently caught in the Mediterranean; and a description is added, by which it may be recognized if ever it should chance to fall into the hands of an observer. We also add a notice of a kindred species which has only been recognized of late as being distinct; but which is enumerated in the Catalogue of the British Museum, where the example is preserved, as having been found by Dr. Leach at Padstow, on the north coast of Cornwall. Our knowledge of this last-mentioned fish is derived from Dr. Gunther, whose account of it therefore, under the name of *Ophidium broussonetii,* we copy; and as these fishes are described as in their form, and in the number of the rays of their fins, closely resembling each other, the plate we give of the one, coupled with a notice of their differences, will be sufficient for every practical purpose.

The usual length of this fish is eight or nine inches, and in shape it may be compared with the Eel or Conger, but that it is stouter in comparison with its length, and also more compressed; the form becoming more slender towards the tail. The body is clothed with small scales of an oblong form, which do not overlap each other. The jaws are equal, and the angle of the mouth a little depressed; gape wide; rows of fine teeth in the jaws, and some in the palate; eye large; lateral line straight. The single dorsal fin, with one hundred and forty rays, or as Risso says, one hundred and twenty, begins over the pectoral and runs to the end of the body, where it becomes united to the anal—forming the tail. Under the throat, and attached to the hyoid bone, is what is strictly a pair of barbles, which, in an example that measured eight inches, were an inch in length; but not far from their origin they are divided into two unequal branches; and this has led to their being often described as four in number. The colour of this fish is variously described, but a prevailing tint on the back is blue; silvery on the sides and belly, sprinkled on the sides with dots. The dorsal and anal fins are narrow and grey with a dark edge.

We have remarked that it is probable this Bearded Ophidium has been confounded with another, which much resembles it, and which stands in the Catalogue of the British Museum as *Ophidium broussonetii*, in honour of a gentleman who wrote a paper on the subject, which is contained in the "Transactions of the Royal Society" for 1781; but it differs in having "only four gill rakers on the lower branch of the outer branchial arch," whereas *O. barbatum* is furnished with five or six. There is also a different form of the air-bladder; an organ the form of which offers specifically distinctive marks in this genus as in many others; and of which, therefore, a figure is given by

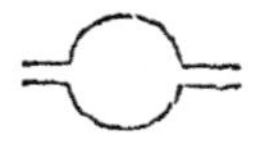

Willoughby as it is found in the Bearded Ophidium. In Broussonet's Ophidium this organ is ovate, without a contracted part; and there is no separate bone which fits into the anterior portion of this air bladder. This species is a native of the Mediterranean.

DRUMMOND'S ECHIODON.

CLVI

FIERASFER.

The body long and tapering; vent placed near the throat. Gill openings wide, the gill membranes united below: particularly distinguished from *Ophidium* by the absence of barbles. An apodal genus of fishes.

DRUMMOND'S ECHIODON.

Fierasfer dentatus,	Cuvier.
Echiodon drummondii,	Thompson; Trans. Zool. Society, vol. ii.
" "	Yarrell; British Fishes, vol. ii, p. 417.
Fierasfer dentatus,	Gunther; Catalogue British Museum, vol. iv, p. 383.

Almost the whole of what we know of this fish is contained in a communication by W. Thompson, Esq., of Belfast, to the Zoological Society, with a figure, which we have copied; and the communication is transferred to the fourth volume of the "Natural History of Ireland" by the same gentleman: to which we add, that an example of the same species has been since caught, or rather found, thrown on the shore by a storm, in the harbour of Valencia, in Ireland; and several others of small size were found by Mr. Edwards at Banff, of which we shall give an account. Mr. Thompson remarks that in external characters it is excluded from the *Ophia* by not having barbles; and although it agrees with the genus *Fierasfer* in being without these appendages, yet by having the dorsal fin elevated and strongly developed, it does not range with them; to which I add, that this character is excluded from our definition as above, since it might seem like a contradiction to classify under such a character the only British fish of the genus, and which cannot be so described The author further says—in Cuvier's Animal Kingdom the *Ophidium dentatum* is described as having in each jaw "deux dents en crochets," but no further details are given: in this only character, however, he judges that the *O. dentatum* differs from the present species, which had four large hooked teeth in the upper, and two in the under jaw. This specimen was found dead on the beach at Carnlough, in the county of Antrim, by Dr. Drummond, in the month of June—thrown on shore probably by a strong easterly wind.

The length of this example was eleven inches; and the greatest depth, which was at one inch and four lines from the snout, was six lines, behind which it became gradually narrower and thinner to the tail. The head was one inch and two lines long—the profile sloping forward equally on both sides to the snout, which is truncated and projects beyond the lower jaw, and is narrow; compressed at the sides, and rather flat above from the eyes backward; from the eyes forward a central bony ridge; a few large punctures extend from the snout below the eye, and are continued just behind it; a series of small ones closely arranged extend from the upper portion of the eye in a curved form posteriorly to near the edge of the preopercle, and thence in a double row extends downwards. Nostrils very large, oval, transverse, a little in advance of the eye. Eye large, occupying the entire upper half of the depth of the head; wider than high, its distance from the snout equal to its diameter; operculum terminating above in a minute point directed backward, and strongly radiated. Mouth cleft a little obliquely. Two large strong teeth placed close together, and curving inwards at each side of the extremity of the upper jaw, the two inmost a little separate. In the lower jaw one slender rounded tooth on each side, curving outward at the base and inward at the point. The upper and lower jaw, and vomer, thickly covered with small bluntish teeth; a series of such teeth on the bones of the palate; those of the upper jaw exposed to view when the mouth is closed. On the dorsal ridge is a short stout spine, the point of which alone is uncovered with the skin. Lateral line scarcely perceptible. Vent one inch and three lines from the end of the lower jaw. The dorsal fin begins at one inch and six lines from the snout; low at first, but becoming wider as it draws near the caudal fin, which it joins. Anal fin begins close behind the vent, is broader than the dorsal throughout, and is widest as it approaches the caudal fin, which it also joins. Near the caudal fin the rays of the anal are wider by four times than the body itself. Middle rays of the pectoral longest. Colour of the first half of the body dull flesh-colour, and behind this are brown markings, as well along the base of the dorsal and anal fins as top of the head, caudal, and hindward rays of the longer fins. Caudal, gill-cover, and a part of the under surface, bright silver. Dorsal fin rays one hundred and eighty, as also the anal; pectoral sixteen, caudal twelve. The vertebræ number eighty-eight. The second example referred to was in length only eight inches.

In the number of the "Zoologist" for April, 1863, was a paragraph which stated that six specimens of this fish had been obtained at Banff, in the preceding March, by Mr. Thomas Edwards, their length varying from four to five inches:—"The teeth most formidable-looking weapons, even in these small specimens. The spine, too, at the back of the head is very conspicuous; no scales discoverable when the fish were fresh from the sea." In consequence of this announcement two of these examples were obligingly sent to me by Mr. Edwards; of which one was laid on a card in a dry condition, and did not measure quite two inches and a half. The other, which was a little longer, was preserved in diluted spirit. The characteristic teeth were plainly seen; and between the two curved prominent ones in front of the lower jaw were others much shorter, and not represented in Mr. Drummond's figure. The mystache long and slender, reaching opposite the hindmost border of the eye. Eye large and oval; no scales. The shape much more slender than in Mr. Drummond's plate, so that Mr. Edwards was inclined to give them the name of Whipfishes. In a letter which accompanied the specimens it is said that when alive the colour was beautifully clear, or of a crystal-like hue, and so transparent that the vertebræ might be counted. The largest example had a fine blush of red down the dorsal ridge, and along the belly, except near the vent, at which part there was an oblong silvery spot; pupil of the eye very dark green, with a white iris; head also white. Their motion through the water was like that of a Wormfish, (*Syngnathus*,) but they never sought to hide themselves among weeds or under stones. They seemed to prefer sandy ground, on which they would lie for hours in a waved or crooked posture. They were obtained near the shore on sandy ground, and not in rough weather; and there seems reason to believe that they were bred near where they were found—a circumstance the more likely, as Mr. Edwards believes that six or seven years before this he found the remains of a large example on the shore near the same place. It measured nine inches in length without the head, which was lost; but the breadth was scarcely more than in the smaller examples, two of which have been transferred to the British Museum. Cuvier notes it as found in the Mediterranean.

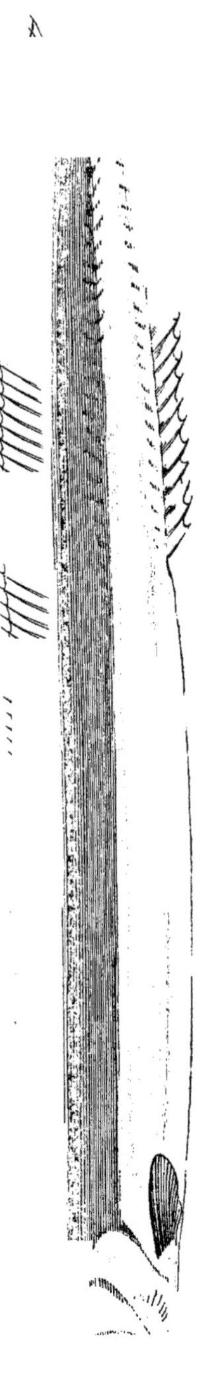

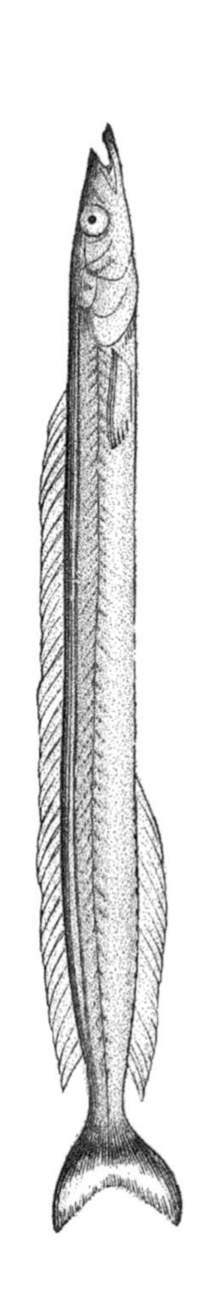

AMMODYTES.

The body lengthened, almost cylindrical, with lines or folds along the length of the body, distinct from a lateral line. Jaws protruded before the eyes, pointed, the lower longest, with a fleshy pointed termination. Dorsal and anal fins long, but not united to the tail. No ventral fins, and consequently a portion of the apodal class of Linnæus: no air bladder.

This little family of fishes obtains its English name from the shape of their body, which at the front ends in a point, and at the tail is so spread out as to form a distant resemblance to the butt-end of the warlike instrument signified by the denomination; while the lengthened and almost cylindrical body answers to the shaft. In some places these fishes are also called Sand Eels, which answers to the Greek word by which the genus is discriminated. They stand as an insulated group, both in regard to form and habits, amongst the native fishes of Britain.

LESSER LAUNCE.

Sand Eel.—Two species are comprised within the name of *Tobianus* by Willoughby, p. 113, and *Ammodytes tobianus* of Linnæus.

Ammodytes tobianus,	Cuvier.
" *lancea*,	Jenyns; Manual, p. 483.
" "	Yarrell; British Fishes, vol. ii, p. 429.
" "	Gunther; Cat. Br. Museum, vol. iv, p. 385.

This little fish is an inhabitant generally of the more sandy shores of the British Islands, and it rarely goes far from them into deep water. They swim in companies of a few scores or hundreds, and in the quiet days of summer are seen at a small depth in the water as they are pursuing their lively course in bays or harbours—where, however, their fate is usually an unfortunate one. Unable to protect themselves, they are an attractive prey to the hungry rovers of the sea, who here and there make a plunge into the midst of them, to the momentary terror of the little host. They are scattered for a time, but they gather closely together again, only, how ever, to be broken in upon by another and another plunge, until at last they find their safety by piercing into the soft sand of the bottom, beneath which the pointed process at the extremity of the under jaw enables them to bury themselves, and in which they lie concealed without injury to themselves, even when the tide has ebbed and left their hiding place uncovered. But it is not only that this fish is able to find its way to shelter in such a remarkable situation; they are able also to move about within it with ease and some degree of quickness; for the better accomplishing of which there appears to be at the root of the tail a special organization, of which the blood vessels are visible, and something corresponding to which exists in all fishes which possess the power of penetrating into the sand or of covering themselves with it. It is in this retreat, concealed and sheltered with the sand of the shore, that this Launce sheds its roe; and this it does as it holds a tortuous course, the grains being scattered as it passes on: and in the west of the kingdom at least this process is accomplished at about the shortest days of the year. It often happens, however, that their hiding place is broken in upon by worse enemies than the prowling natives of the deep; and people who value them as a delicacy resort to their retreat with hooks or rakes, and thus draw them up to light. I have been informed by those who have been accustomed to this practice, that if the Launce be touched with the hook on the posterior part of its body, it will move away through the sand with such celerity as scarcely to be again overtaken; so that it requires some skill to succeed in what might appear so easy an employment as raking these fishes out of the concealment of the sand.

In some places the Lesser Launce is a favourite bait with fishermen; from some of whom I learn further, that when Mackarel are discovered to be in pursuit of the Larger or Wide-mouthed Launce, a less successful fishery is expected; but when these lesser fishes are the object of their rapacity, the fishery shews itself much more profitable.

The range of this fish extends itself to the coasts of Sweden and Norway, and even much farther to the north; but it is not numbered among the fishes of Madeira by Mr. Lowe; nor do I find it mentioned by such writers on the Natural History of the Mediterranean as I have been able to examine, for it is now known to be a different species from the Larger Launce which has been described by the more ancient authors; and it may also differ from *A. ciceretus* of Rafinesque, the name of which is also mentioned by Willoughby, but which has not yet been distinctly characterized. This Lesser or Sand Launce usually grows to the length of four or five inches, but its form is best described by comparison with the Large or Wide-mouthed species, which will come next under our consideration

LARGER LAUNCE.

WIDE-MOUTHED LAUNCE.

Tobianus,	WILLOUGHBY; well described, but with reference to a figure of another fish.
"	JAGO; in Ray's Synopsis, f. 12.
Ammodytes tobianus,	LINNÆUS. CUVIER.
" "	FLEMING; British Animals, p. 201.
" "	JENYNS; Manual, p. 483.
"	YARRELL; British Fishes, vol. ii, p. 424.
" *appat,*	RISSO. BLOCH; pl. 75. DONOVAN; pl. 33.
" *lanceolatus,*	GUNTHER; Catalogue British Museum, vol. iv, p. 384.

THIS is a fish of great activity, as might be judged from its slender form, and well-constructed shape; and it is also voracious, so that it pursues and devours some which might be supposed little likely to become its prey. Its gape also is capacious, by the aid of which it has been able to swallow the hook, baited with a lask or slice cut from the shining side of a Mackarel, and which was intended to have proved an attraction to a much larger prize. It has happened not unfrequently that the Lesser Launce, which formerly was believed to be in the half grown condition of its own race, has been found in its stomach.

The favourite resort of this species is in much deeper water than is frequented by the Lesser Launce, and thus it has been known to have been devoured by the larger fishes which have been caught at ten leagues from land, at the entrance of the British Channel, in a depth of forty-five fathoms; and this, too, in the middle of the summer, although that is the season when it is common for it to draw near the land; in doing which it may be said to perform a regular migration. Risso represents these fish as appearing thus in the neighbourhood of Nice; where they are seen coming in schools from the west in May and June in their passage eastward; at which time, although of small size, they are fished for with nets, which also take the smaller sardines and anchovies (*clupeidæ.*)

With us these Larger Launce are in much more limited numbers than the smaller species, and more locally as well as sparingly distributed: and I have no knowledge of any district in which they form a particular object of interest to fishermen except on the smooth sandy shore of St. Ives, on the north coast of Cornwall. The net prepared for this fishery is about twenty fathoms in length. and of a depth to suit that of the water; the meshes being of the size barely to admit the passage of a sixpence, and at the middle of the net, where it is formed into a bag, there is what is termed the bunt, which consists of a fine sort of canvas. The boat is of good size, and has a crew of four men; with another man who stands on the shore and who holds fast the warp that is fastened to the end of the net. The boat is by this means kept in the proper position with her broadside towards the shore, and thus the net is made to form a circle, so as to enclose the fish within it until it is drawn up or tucked into the boat, with the fish collected together in the bag. It must be owing to the nature of the ground, which consists of the finest sand, with scarcely a stone to interfere with its smoothness, that it is not usual for other fishes, even of small size, to be enclosed in the net with the Launces; but the numbers of the latter taken at one haul will often amount to a couple of bushels; and even three bushels have been enclosed in the net at one time.

This fishery usually begins in May, and will continue until September, if the more profitable fishery for Pilchards does not intervene to put an end to it. The Launces thus caught are chiefly employed as bait for the larger fishes; and without them the hook fishery in that neighbourhood would be greatly interfered with, if not destroyed—for in that district there are few of the resources which abound in other parts of Cornwall to supply fishermen with bait. But these Wide-mouthed Launces are also sold for the table, and where they are known they are represented as being a great delicacy. This species is mentioned by Nilsson as known along the coasts of Scandinavia, and it is also met with, as we have seen, in the Mediterranean; but it is not mentioned by Lowe among the fishes of Madeira, although it is reported as having been obtained on the coasts of America

The usual length of this species is about a foot, but the measure assigned by Jago to the specimen of which he has given a figure, was almost sixteen inches, and an example of that full length is preserved in the British Museum. Its general form is lengthened, and only a little compressed at the sides; the shape continuing uniformly from the head for about three fourths of the length, but becoming gradually more slender towards the tail. The head in front of the eye tends to a point, which when the mouth is shut ends in the lower jaw, where there is a projecting cartilage by which this fish is able to pierce its way into the sand. The gape is wide, and is rendered the more so by the faculty it possesses of lifting the upper jaw into a right angle with the front of the skull. The mystache is wide and reaches far back, but scarcely to the front border of the eye. No teeth in the jaws, but there are some in front of the palate. Several specks like perforations on the head. Eyes conspicuous; the gill covers project backward above the root of the pectoral fins. There are five longitudinal lines on each side, one of which is near the back; another is the true lateral line, and is marked with the insertion of muscles; and three are on the belly, of which two appear to unite at the vent. This orifice is about two thirds of the whole length from the snout, excluding the caudal fin. There are no scales, but the skin is marked with oblique transverse folds, which, according to Dr. Gunther, are one hundred and seventy in number. Pectoral fin moderate and low down; the dorsal begins a little behind the termination of the pectoral, and passes at an uniform height to a little short of the tail; as does the anal from near the vent. Tail concave with rounded borders. Colour of the top of the head and back, down the sides to the true lateral line blue; tail bluish: all below, with the cheeks, brilliant white; dorsal and anal fins pale white. At the root of the tail, during life, there is a circulating vessel, which appears to have some connection with the faculty of burying in the sand, since something of the same nature is perceptible in other fishes which are endued with a similar propensity.

Compared with the Lesser Launce this last named fish has the gape proportionally smaller, the mystache less extended; palate without teeth; eye nearer the angle of the mouth. The dorsal fin begins nearer the head and considerably in front of the termination of the pectoral fin. Dr. Gunther counted the oblique folds of the skin as from one hundred and twenty to thirty in number. The number of the fin rays is in both these fishes nearly the same.

SHORT-SNOUTED LAUNCE.

Ammodytes cicerelus, RAFINESQUE. SCHMALTZ?

IT has been long supposed that besides the two kinds of Launces we have already described there might be one, or perhaps two, more to be found in the seas of Europe; and accordingly the wave-finned species (*A. siculus* of the Catalogue of the British Museum,) has since been distinguished by Mr. Swainson as an inhabitant of the Mediterranean. But there is another which is briefly described by Rafinesque under the name given above, and by which it appears to be known to Italian fishermen, although it has hitherto been confounded with the Wide-mouthed Launce; and even Cuvier appears to have been disposed to admit it as a distinct species, although he has generally shewn a reluctance to admit the accuracy of the Italian writer above quoted in his descriptions or arrangement.

The foundation of the belief that the *Cicerelus* is probably a native of the British seas, is so far deficient that it rests on the examination of a single specimen; but I have reason to believe that this is to be ascribed to the want of opportunity rather than to the absolute rarity of the fish; and as it was so long confounded with the better known species by observers in the Mediterranean it should not excite surprise to find that it has been equally mistaken for the other in Britain.

The specific characters assigned to this fish by Rafinesque are—that besides the colours—which of the back are bright blue and of the under parts silvery white, the eyes are situated immediately above the corner of the mouth, and (which is less decisive,) the dorsal fin rises a little behind the line of termination of the pectoral. The chief objection which will be felt in this instance arises from the fact that the Italian writer describes the *Cicerelus* as rarely exceeding the length of

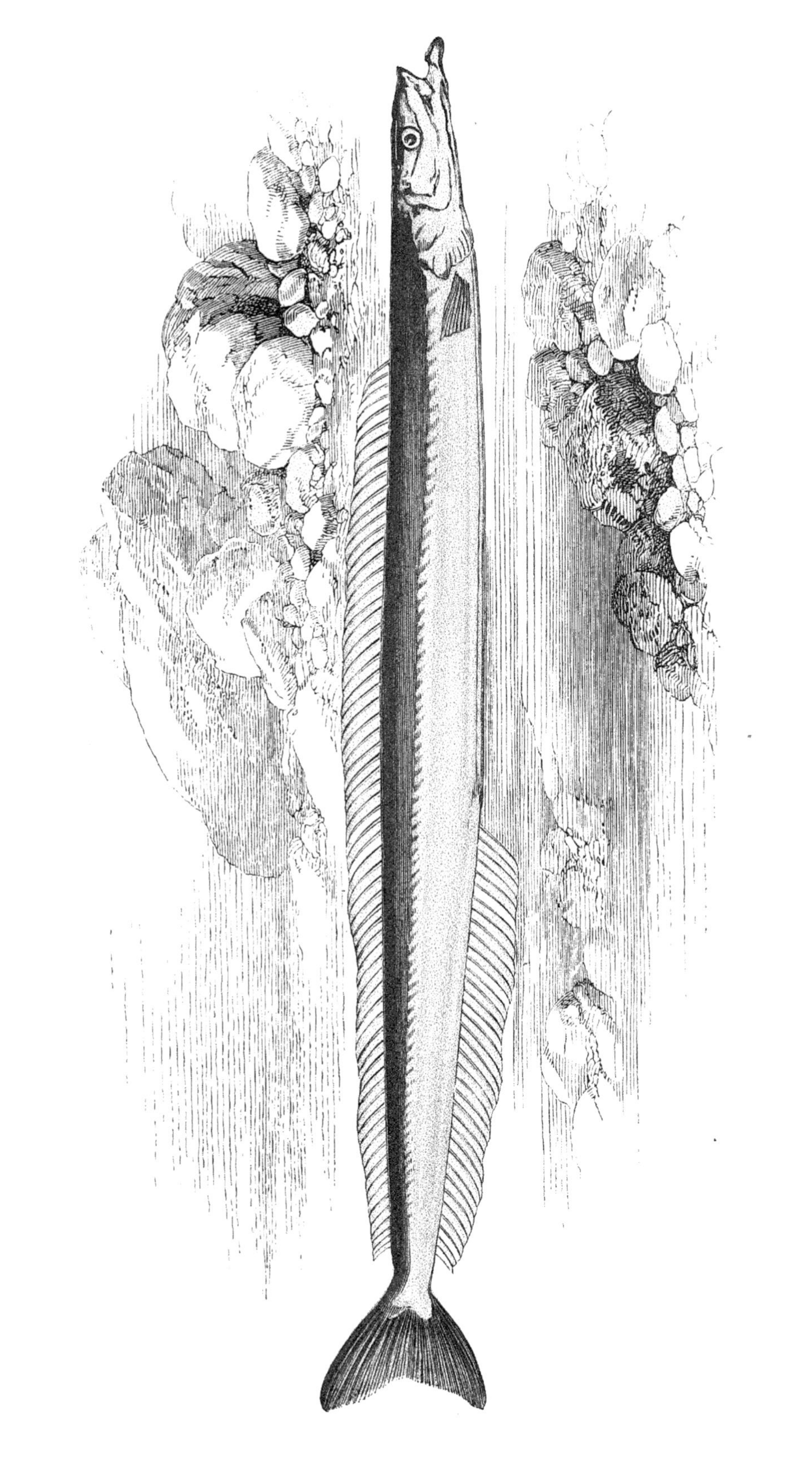

three or four inches, while that one which came under my notice measured a little more than eleven inches and a half; but on the other hand he so far departs from his own assigned length as to give a figure which measures a little more than five inches and a half; and so confined has been his observation of these fishes, that no other species finds a place in his book.

Compared with a Wide-mouthed Launce of about the same size, the shape of this fish was rather more stout and angular, and this stoutness was carried nearer to the tail, so that it became contracted more suddenly between the termination of the dorsal and anal fins and the tail. At the union of the body with the caudal fin also there was more decidedly a heart-shaped termination than in the Wide-mouthed species. But a more distinguishing difference between the two was in the head, before the eyes; this portion being much less thrust forward, so that the point of the upper jaw was more rounded, and to a less extent capable of being bent up, and the gape much less,—the mystache did not extend above one third of the space toward the eye; and although indeed the eye was not situated immediately above the angle of the mouth, as expressed in Rafinesque's specific character, it stood nearer to that point than in the Wide-mouthed species. The point of the lower jaw also was much less protruded. Along the belly there ran seven longitudinal lines, of which those on the median line were close together, and the two next were bordered with a slender membrane, which may be of some use as organs of sensation. The oblique lines on the sides were finely grained, but their number was not counted. The dorsal and anal fins appeared wider than in the more common species. Colour of the back and top of the head fine blue, which does not come down to the lateral line, as I have usually observed it in the Wide-mouthed Launce; sides, belly, and cheeks silvery: the longitudinal line next above the ventral line whiter than the rest of this surface: the tail more expanded than in the other species, with a white border on its under edge.

This example came to me on the first of December, and was distended with roe.

PLEURONECTIDÆ.

THE FAMILY OF FLAT FISHES.

A TOTAL want of uniformity between the sides of the head, and also generally of the body. Both of the eyes are on one side of the head. while the nostrils generally maintain their position in pairs on each side of the summit The mouth is twisted, so that the two corners do not answer to each other, and within the mouth there is generally a veil or cross membrane above and below. The body is much compressed: one side bearing colour, the other without it, and the structure of the skin differs in each. The abdominal cavity very short, but the lobes of roe are contained for the most part in a recess which passes backward between the muscles towards the tail. The vent close to or enclosed between the ventral fins, which are near the throat. The fins nearly encircle the body, and the dorsal fin begins on the top of the head or in front of it.

In most if not all of these fishes the spinous processes of the vertebræ which support the rays of the dorsal fin, and are anterior to the pelvis, are turned forward for that purpose; while those of the vertebræ on the same line of the back that are near the tail are directed backward. On the abdominal side all the spinous processes behind the abdomen itself are directed more or less backward. It is an essential part of the structure of these fishes that the superior and inferior processes of the vertebræ are very long, and those of the sides very short; but in the Carter, and perhaps in others of this family which possess a lengthened form, while the superior and inferior processes are comparatively shorter, the lateral processes are somewhat more developed. In the Topknot the spinous processes of the vertebræ are remarkably long, and anterior to the first of these processes there are numerous (what we term) intermediate bones for the purpose of sustaining fin rays, and which are continued to the snout—in their course resting on a thin ridge of bone which forms a crest on the head. The border of the abdomen of this fish, from the first long pinons process of the vertebræ to the throat, is bounded by a curved bone, the concavity of which supports the entrails; and along its convexity pass off the intermediate bones of the fins. As an example of some of the differences between the eyed and blind side of these fishes, we notice of the Flounder that on the coloured side the masseter muscle, which closes the mouth, is strong, and sends a stout tendon to the angle of the lower jaw, while on the white side there is no tendon, but a well-marked nerve crosses obliquely from the head to the angle of the jaw. A similar nerve on the coloured side descends under the tendon, but comes out of the skull further forward, and passes close below, or on the facial side of the eye. A blood vessel accompanies this nerve in both instances. The nerve running to the palate appears to be the largest in the body, and passes only on the coloured side.

The *Pleuronectidæ* are so called from their habit of swimming flat on the side, (the lateral line of the coloured part being their upper portion,) and not erect, as is the case with other fishes. This curious mode of action is associated, as we see above, with as remarkable a variation of shape and inward structure, when compared with their fellow inhabitants of the water; so that we may confidently pronounce them to be the most irregular and strange of all the creatures we meet with, and an exception especially to other vertebrated animals. We may add, that if they had been met with only in a fossil condition, there is reason to believe they would have been set down as having found their place on earth when the beings inhabiting it were struggling to pass from an unformed to a regularly organized state, without its being certain in what at last the struggle might end. And yet this family of fishes are perfect in themselves, and their form is well adapted to the peculiarities of their modes of life; with the reversal, however, of the functions of all their fins; for even that of the tail, while it serves, as in other fishes, as an organ of propulsion, is still made to act horizontally, and not laterally; and as thus it is of little use in turning the body in its course, the pectoral fin for the most part is brought into action for this purpose.

In separating this numerous family into subordinate groups, it is convenient and usual to divide them principally according as when the belly is towards the observer the head and eyes are directed to the right or left. But this particular structure is not known to exert any influence on the habits of the individual species; and, therefore, in combination with this it is of importance to include in our consideration other material variations of form; which, therefore, are placed among the essentials of the definitions of the genera.

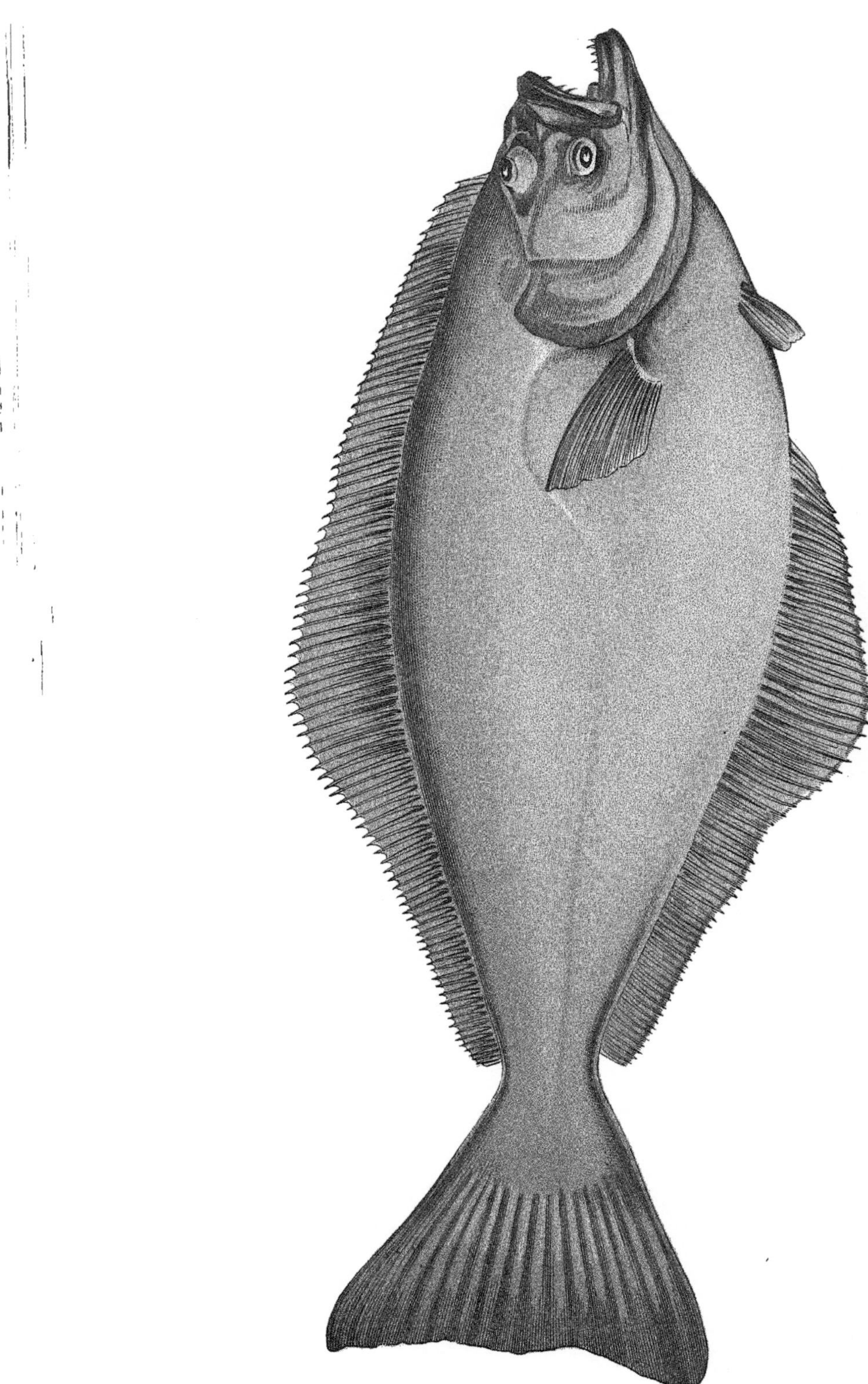

HIPPOGLOSSUS.

When the vent lies toward the observer, the eyes are directed toward the right; a double row of teeth in the upper jaw; no teeth in the palate; gape wide. The dorsal fin begins above the eye; lateral line arched at its beginning.

HOLIBUT.

LADYFLUKE.

Hippoglossus,	Willoughby; p. 99, pl. f. 6.
Pleuronectes hippoglossus,	Linnæus. Donovan; pl. 75.
Pleuronecte fletan,	Lacepede.
Hippoglossus vulgaris,	Fleming; British Animals, p. 199.
" "	Jenyns; Manual, p. 460.
" "	Yarrell; British Fishes, vol. ii., p. 323.
	Gunther; Catalogue Br. Museum, vol. iv., p. 403.

This species is a native of the deeper water which lies to the north of Europe and America, but it wanders to the south so far as to be met with at times on all the coasts of the British Islands; where, however, it is scarcely sought after, nor is it regarded as a very valuable prize when caught. When of moderate size indeed, as three or four feet in length, they are palatable food, but when of the magnitude they often reach they are not esteemed for the table with us—except perhaps the head, which is represented as being a delicacy. But in Greenland, where it is common, this fish is greatly valued; and this, perhaps, for the very properties which cause it to be generally banished from fashionable tables in Britain; and by fishermen of those northern regions a regular arrangement is practised for procuring it. This we learn from Lacepede, who particularly mentions what may be termed the boultey, such as we have already described when we spoke of the manner in which our own fishery is conducted for catching the Ling and Cod; and by this means he says that three or four of the Holibuts are taken at one haul. As it appears that they give a preference to the same ground with the Cod—probably from feeding on a similar food with that fish—we may conclude that the capture of the last named fish is not less an object of attention in this method of fishing than the Holibut itself.

According to Bloch these fishes may be said to associate together, although this attraction may proceed only from the predacious habits which they have in common, since he represents them as lying at the bottom in rows with open mouths, waiting for the approach of fishes or crabs that may come within their reach; and if unsuccessful in their object, it is said that their hunger then urges them to make a formidable assault on the tail of their nearest neighbour. If we may adopt this explanation, and apply it to other members of the same family of fishes, we shall be able to account for the frequent injury which we discover in the tails of some sorts of Flatfishes, as seen especially in the younger stages of their growth; but it may also be caused by the more insidious depredations of other prowling animals which inhabit the bottom of the sea.

It appears that at times this assembling of Holibuts is on a bank of sand in shallow water, where they lie basking in the sun; and when discovered they are assailed with the spear, with which the fishermen endeavour to pierce and fix them to the ground; but in doing this much skill, strength, and patience are necessary to prevent the boat from being swamped or overturned by the powerful struggles of the fish, which are known to be very violent, and as such are recognised by British fishermen when the Holibut has chanced to swallow their hook. By them this fish is represented as being very wild, and its powerful efforts to break loose are characterized by violent and sudden jerks in various directions. When, therefore, it is pierced with the spear, the fishermen of the north proceed to raise it very slowly, and when brought within their reach a club is brought into action, by which, as soon as possible, it is deprived of life.

Along the coast of Norway the fishery for the Holibut is

followed only when the spring is well advanced and the nights are clear; at which time the fish are discovered as they lie at the bottom, even where the water is deep; but as summer advances the search is given up, because, as this fish becomes very fat, the heat of the weather would prevent its flesh from being properly dried in the manner they are accustomed to prepare it. Each portion of this dried fish has its separate name among the people, of whose subsistence, as well as of their commerce, it forms a considerable part. The skin, and even the entrails are also regarded as of value by a people who, from the scarcity or cost of other materials are led to employ these parts for purposes which in more favoured regions were better answered by other means. Perhaps there is no portion of the ocean in which the Holibut is more abundant than on the banks of Newfoundland, and it is there we have heard of its being taken of much larger bulk than is usual elsewhere; perhaps from the abundance of congenial food, for which its appetite is represented as being very keen, as its power of swallowing is also great; so that it is able to gulp down fishes of considerable size, as well as crabs and shellfish. It spawns in the spring.

The Holibut is by far the largest of the Flatfishes; so that in some rare instances it seems scarcely an exaggeration in Lacepede to compare it in size with a whale; especially if we are to suppose the comparison to be made with reference to other species of the same family. The largest I have seen weighed no more than one hundred and twenty-four pounds; but Pennant mentions an example that weighed three hundred pounds, and one which was caught near the Isle of Man in April, 1828, measured seven feet and six inches in length, with the breadth of three feet and six inches, and in weight was three hundred and twenty pounds. This, however, is little more than half the weight of one that is reported to have been taken in Iceland; but which again must have been considerably less than an example mentioned by Olafson, which measured but little short of twenty feet in length. I have also been informed by an officer of the navy that he was present at the capture, on the banks of Newfoundland, of a Holibut which greatly exceeded in size even the example mentioned by Olafson.

The general shape and proportions of this fish will be understood by the comparative measurements given above. The capacity of the mouth is large, and the jaws are armed with strong and prominent teeth, those in the lower jaw stout, scarcely close together, and incurved; in the upper jaw with an approach to a double row of irregular size. The right eye a little in advance of the left; nostrils on the ridge of the head. The body smooth, but when dry the scales appear oblong and overlapping each other. Lateral line curved over the pectoral fin, afterwards straight. The dorsal fin begins over the eye, at first narrow, expanding at about two thirds of its length, afterwards narrower, ending short of the tail: anal fin beginning at a considerable distance from the ventrals; expanding at an early portion of its progress, narrower towards its termination; rays of the former one hundred and two, of the anal seventy-four, but prone to vary in number. In the pectoral I have found fourteen rays in the coloured side and fifteen below, but this is probably liable to variation. Caudal slightly incurved, the border waved, sixteen rays; ventrals small, distant from the throat and also from the anal fin. Colour of the upper surface livid, with perhaps a tendency to olive, below white.

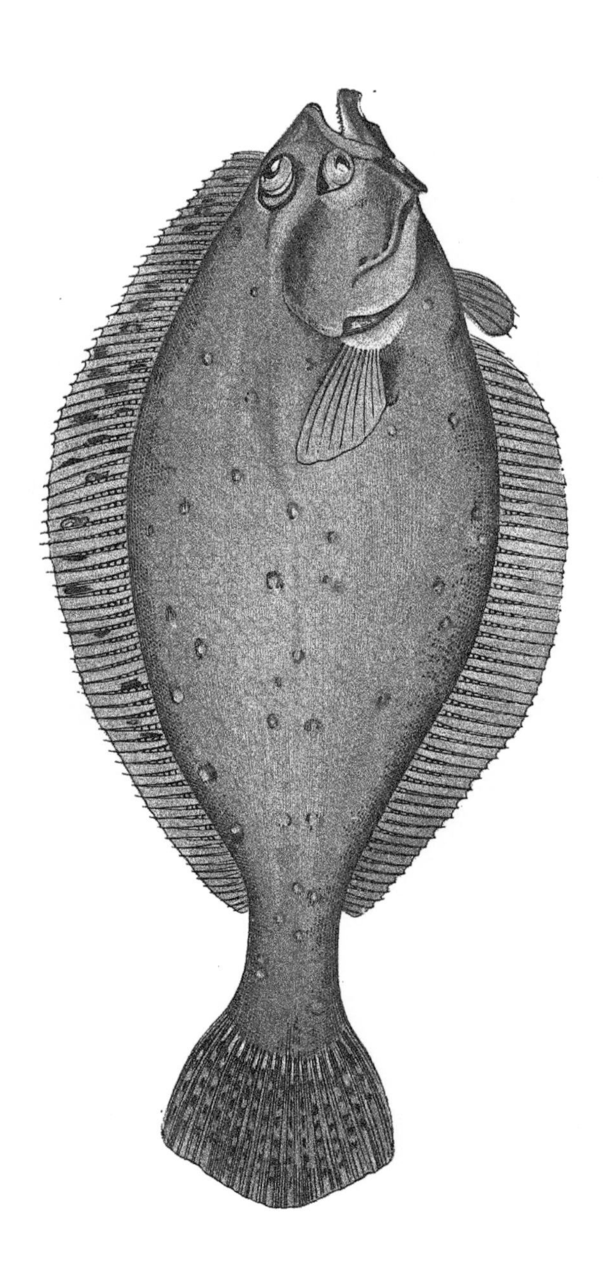

HIPPOGLOSSOIDES—GUNTHER.

EYES directed towards the right; mouth wide, teeth small, in a single row; dorsal fin beginning above the eye; lateral line without a curve at its origin; (ventral fins neither close to the throat nor to the anal.)

LONG ROUGH DAB.

SANDNECKER. LONG FLEUK. ROUGH FLOUNDER.

Citharus flavus sive asperus,	RONDELETIUS.
Pleuronectes limandoides,	BLOCH; pl. 186. TURTON'S Linnæus.
Pleuronecte limandoide,	LACEPEDE.
Platessa limandoides,	JENYNS; Manual, p. 459. YARRELL; Br. Fishes, vol. ii, p. 313.
Hippoglossoides limandoides,	GUNTHER; Catalogue Br. Museum, vol. iv, p. 405.

THIS fish is scarce and but little known in Britain; so that when obtained it is thought worthy of particular notice. It is especially a native of the north, and is remarked as being taken more commonly than elsewhere near the island of Heligoland; but it is also enumerated among the fishes of Norway and Sweden, although it does not exist in the Baltic. Southward its range is limited; but it is reported by William P. Cocks, Esq., as having been purchased in the fish-market at Falmouth. Although we suppose it to have been known to Rondeletius, it is not mentioned as having been met with in the Mediterranean nor anywhere further south. It is said to prefer sandy ground, at a great depth, and its food, as in the generality of this family, is taken from the bottom. Mr. Cocks found the stomach of his specimen filled with the shells of *Turritella terebra,* and two thirds of the number contained hermit crabs—*Pagurus lævis.* It spawns in May and June.

For the sake of greater accuracy, I copy our description and figure from the Swedish work of Ekstrom and Sundevall; of which it is the boast that the plates and descriptions were taken from the fishes represented in it immediately as they came out of the water. Mr. Cocks' example was caught in the middle of September, and measured nine inches in length, and in breadth three inches and a quarter. According to Ekstrom, it sometimes reaches twelve inches, with the breadth, exclusive of the fins, one third of the extreme length; the head to the border of the gill-cover is one fourth of the whole body, excluding the caudal fin; the body regularly oblong, much compressed; in thickness not exceeding a tenth of the breadth; the scales more regular and equal than in most of the flatfishes—rounded, with a free border, doubly bent or channeled, the middle angle blunt, rounded with from twelve to twenty small points, which are ciliated; on the blind side these are so roughened only at the hindmost part and at the base of the fins. The eyes, which look to the right, are nearly equal, and the upper one is placed by a fifth part further back than the other; the lower eye in length equal to the distance of the point of the snout. The ridge behind the eyes low and smooth. Mouth large, as in the Holibut and Turbot; teeth small, conic; the usual veil in the mouth as in all flatfishes. Lateral line only gently sloped; vent near the angle of the gill-cover; anal spine strong and sharp. Dorsal fin with from seventy-eight to eighty-five rays, and in breadth at the middle one fourth of the breadth of the body; all the rays with scales in a single row, which are rough on the upper side; the anal fin begins a little behind the vent, and has from sixty-four to sixty-six rays; this and the dorsal ending far short of the tail; caudal fin lengthened at the middle, with eighteen rays; pectoral with ten or eleven rays, which are all simple—a circumstance that is peculiar to this species; but the lower pectoral, which is smaller, has branched rays; ventral fins with six rays. Colour dusky yellow, and in some situations it is slightly spotted.

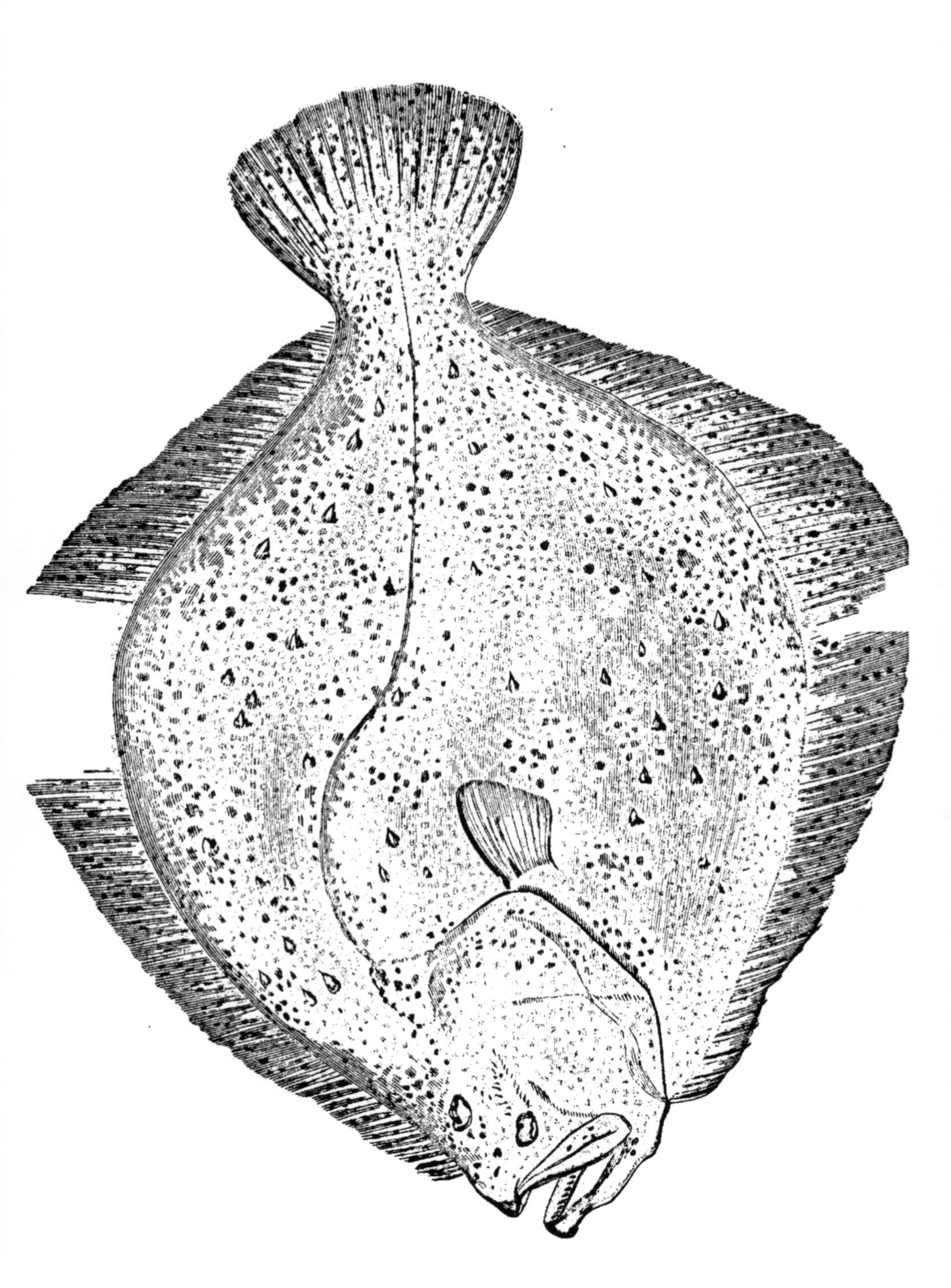

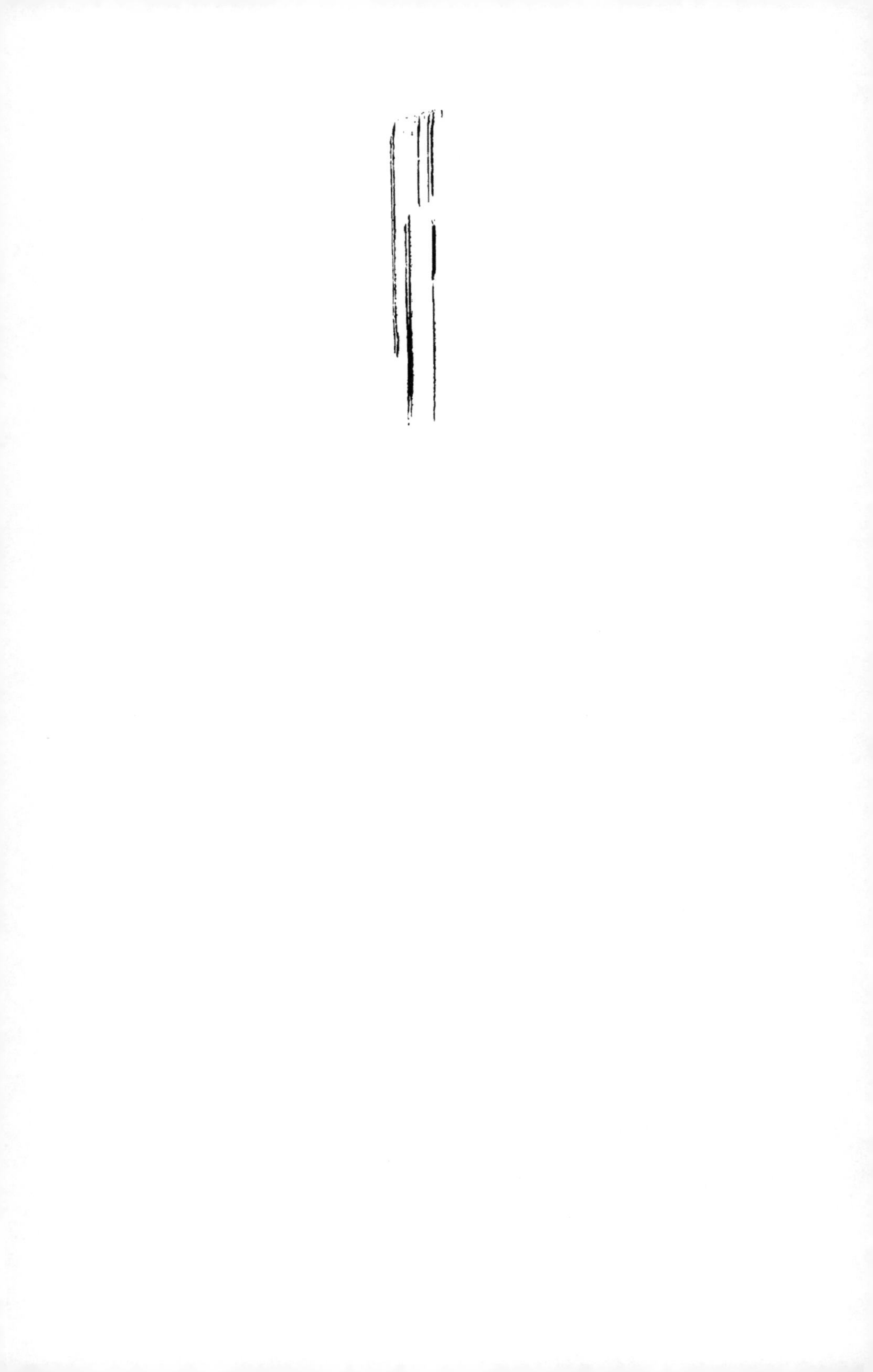

RHOMBUS.

The eyes are directed towards the left; gape wide, the jaws with a narrow band of teeth, and some in the middle of the palate (vomer;) dorsal fin beginning close to the upper jaw; ventral fins near the throat, and also near the anal.

TURBOT.

BANNOCK FLEUK. TARBUT.

Rhombus aculeatus,	Jonston; pl. 22, f. 12.
" "	Willoughby; p. 94, pl. f. 2; but the dorsal fin is not placed sufficiently near to the lips.
Pleuronectes maximus,	Linnæus. Bloch; pl. 49.
" "	Donovan; pl. 46.
Pleuronecte turbot,	Lacepede.
Rhombus maximus,	Cuvier. Fleming; Br. Animals, p. 196.
" "	Jenyns; Manual, p. 461.
" "	Yarrell; Br. Fishes, vol ii, p. 324.
	Gunther; Cat. Br. M., vol. iv, p. 407.

The Turbot is one of the best known of this family of fishes in consequence of the estimation in which it is held at the most luxurious tables, where it is equalled only by the Salmon and Sturgeon; and the same high value was set on it in the prosperous days of ancient Rome, as we learn from several ancient writers, who inform us it was called the Sea Pheasant, and who mention some of the extravagances to which a compliance with the fashion led its followers. Horace says,

Great Turbots and the soup-dish lead
To shame at last, and want of bread;

but, contrary to the present taste, the preference was given to

Turbots of the largest size; and Martial refers to a feast where

> ——the enormous fish
> Was wider than the broadest dish.

A more remarkable circumstance connected with its ancient history, as significant of foolish despotism on the one hand, and on the other of the degradation of a once illustrious assembly, is that in which the Emperor Domitian is said to have summoned a meeting of the Senate, or rather they assembled of their own accord, when the object of the meeting was found to be—that they might advise the sovereign what should be the sort and size of the vessel in which might be cooked a mighty Turbot that had been brought to him. It must be confessed, however, that this oft-repeated story appears to stand in need of confirmation, since, while it is mentioned by a satirist, it is not referred to by Suetonius, who has shewn no reserve in speaking of the bad deeds of this prince; and who, if it were true, cannot be believed to have been insensible to the insult thus offered to, or the disgrace incurred by the nobility of the empire.

For the more ready supply of this much-coveted delicacy, Turbots were preserved in ponds of salt-water; and this seems to have been the more necessary, since there is reason to believe that this fish is not very generally distributed in the Mediterranean. Dr. Gulia, in his account of the fishes known at Malta, regards it as of casual and rare occurrence in that island; and it is not named by Rafinesque among the fishes of Palermo.

Attempts have been made to prove that the fish referred to in the satirical poetry of Juvenal, as also in other ancient authors, under the name of Rhombus, and among the Greeks as Psetta, was not the Turbot, but the Brill, another species which comes next in the order of our enumeration. There is much probability that these fishes, which nearly resemble each other, were often by the ancients confounded together; but the particular reason why the Brill has been judged the more likely to have been the true Rhombus of antiquity appears to be that there were a few fishes, among which were the Rhombus and Asellus, which were believed to conceal themselves in the sand or mud at the bottom, and there spread out the rays of their fins, and move them to and fro to attract little fishes so as to devour them. It is further said that the true Turbot does not possess such rays to its dorsal fin as are fitted to this alleged use, while the Brill is thus furnished, as indeed may be seen in a sketch we give of a fish of that species which is thus ornamented or supplied.

It is certain, however, that there is a mistake in these particulars, and consequently in the argument which is derived from them; and that this development of the first rays of the dorsal fin is not only *not* a circumstance peculiar to the Brill, but when it occurs that it is only an exceptional case in that fish—occurring only once in many instances, without reference to sex or age; while it is found just as often in some others of the same family.

In one example of a Turbot, in which both sides were alike in colour, there proceeded from before the eyes a process three inches in length, thin in substance, and nearly as wide as the breadth of the finger. It was directed forward, and so far differed from the rays of the fin as not to be connected with them; but although this may have been an abnormal formation, yet an enlarged development of the front rays of the dorsal fin is in this fish far from being an extraordinary occurrence. In this argument, so far as regards magnitude, an objection lies against the Brill, that it is never found of equal size with a Turbot of full growth; and the following instances tend to shew that in some cases the Turbot only can answer the requirements of Juvenal's sarcasm or Martial's epigram.

It is not uncommon to meet with one of the weight of thirty pounds, and I possess a note of an example of which the weight was seventy pounds; but even this was exceeded by one which is mentioned by Lacepede; another is recorded to have been caught in Scotland of the weight of ninety pounds and a quarter; and these again were far exceeded by one which that eminent naturalist, Rondeletius, informs us he himself saw, which in length measured five cubits, or seven feet and a half, with a breadth of four cubits, and in thickness a foot; which dimensions will fully answer, and perhaps exceed all that is said of the Turbot of Domitian, but which cannot be applied to any Brill of which there exists a record. Again, in the account given of this transaction, one of the speakers refers to the *sudis*, or spikes, which were on the back of the fish, an expression not correct in reference to the fin rays of either of these fishes, but strictly applicable to the upper surface, or what fishermen term the back, of the Turbot. It is to be observed further that this presentation to the emperor of an immense Adriatic Turbot was scarcely a voluntary act, since informers would have been ready to carry the news to the prince, and thus have ruined the fisherman.

The Turbot is a fish of northern or temperate climates, and is said to grow to a larger size generally on the coasts of Britain and France than further south; but it is also known along the shores of Italy and Greece, and it is found also in the Black Sea. It prefers sandy ground, or where there is gravel; and it is also reported to choose a bottom of mud, in which to embed itself for the purpose of hiding its body, in order the better to entrap unwary fishes; but this faculty of intelligence will require more positive support from observation than it has yet received. It appears to wander much, and in small companies; and I have been informed by fishermen that in many instances, when one has been drawn up with a line, a companion has followed it so closely as to be taken with the aid of the usual hooked stick (gaff) employed in lifting on board the larger fishes. But although the usual habit of the Turbot is to lie close to the ground, it is seen to mount occasionally to the surface, and maintain its station there at one stay for a considerable time, as if enjoying the flowing of the current; but in that situation it has seemed less eager to take a bait.

The ocean north of the Straits of Dover is a favourite resort of these fishes; and it is there more particularly that a regular fishery with long lines, or bulteys, is carried on for catching them, the bait being a portion of a Herring or a Lamprey, large quantities of this last-named fish being collected for sale to the Dutch for this purpose. Mackarel is also a favourite bait, but only for a short time in the season. The fish which are thus taken are brought for the most part to the London market; and in the middle of the last (eighteenth) century, the fishermen of Holland are said to have received one hundred thousand pounds in one year for what they had brought to England of these fishes.

It is requisite to successful fishery for Turbots with a line that the bait shall be newly killed; and a living bait is still more attractive, for this fish is not a little ravenous, and if it chance to escape from the hook, it will again and again encounter the same risk. On one occasion a Grey Gurnard had swallowed the bait, when it was itself seized by a Turbot, which, in passing it into its stomach, head foremost, suffered the mischance of having the spines of its prey to become fixed in its gullet, so that both of these fishes were drawn up together. Crabs and shell-fish also form part of its food, and indeed it appears that little which has life is rejected.

The trawl in the west and south of England is extensively used for the taking of Turbots, as it is indeed for obtaining every sort of fish that falls within the sweep of its net; but more especially it is successful for those of the *Pleuronectidæ*. But the fishes are bruised, and for the most part greatly injured in this method of fishing,—as may be imagined, when we call to mind that they are dragged along on the ground for a considerable distance, amidst an accumulation of whatever heavy substances may come in the way. This fish is retentive of life, so that it will remain a whole day alive after being caught; and yet when brought to Billingsgate they are sometimes so much decayed as to be unfit for food.

By an Act of Parliament, (1st. George I., C. 28,) a Turbot is forbidden to be sold when under the length of sixteen inches, Brill or Pearl fourteen inches, Codling twelve, Whiting six, Bass and Mullet twelve, Sole eight, Plaise or Dab six, Flounder seven; but there is no penalty for catching them of less size than is here specified, and consequently the prohibition itself affords no advantage towards what appears to have been intended by it.

The breadth of the body of this fish is contained once and three fifths in the whole length, excluding the tail fin; and consequently it is wider proportionally than any other of the British flatfishes, except those much smaller species the Topknots; and from this greater breadth and more rounded form it has received in Scotland the name of Bannock, or Cake Fluke. The gape is wide, opening obliquely downward, with a mystache which reaches opposite the anterior eye; the eyes separate, the lowermost a little in advance; a flat projecting spine covering a portion of the upper jaw; under jaw longest; teeth placed thickly in several rows, and a row also in the palate. Lateral line arched over the pectoral fin, and thence straight to the tail; upper or coloured side without scales, but irregularly covered with stout bony tubercles; on the mystache, round the eyes and the gill-covers, are thickly-set small bossed points. The tubercles on the body have not yet appeared in very young examples, and they appear to be less in number in examples which seemed to be of advanced age. The dorsal fin begins in front of the eyes, and, with the anal, passes on close to the tail; the tail round. Ventral fins close under the throat, and nearly joined to the anal. The hindmost nasal orifice on the blind side, bordered posteriorly with a loose membrane. The colour varies in intensity, from light brown to almost black; and in an example that was intensely black, both sides were alike coloured, mottled and sprinkled over with dots of darker and white, as if with sand. The fin rays are wide: in the dorsal sixty-eight; anal forty-eight to fifty-nine; pectoral ten; caudal fifteen to seventeen; ventrals five on the coloured side, six below, the first of the latter answering to the second above.

BRILL.

PEARL. KITE. LUGALEE, OR LUGALEEF.

Rhombus non aculeatus squamosus,	WILLOUGHBY; pl. f. 3—the under side.
Pleuronectes rhombus,	LINNÆUS; RISSO.
" "	FLEMING; Br. Animals, p. 196.
" "	JENYNS; Manual, p. 462.
Pleuronecte carrelet,	LACEPEDE.
Rhombus vulgaris, Barbue,	CUVIER.
" " "	YARRELL; Br. Fishes, vol ii, p. 331.
Rhombus lævis,	GUNTHER; Cat. Br. Museum, vol. iv, p. 400.

THE *Pleuronectes cyclops* of Donovan appears to me to be the Brill a little deformed about the head. Dr. Gunther supposes it to be the young state of the Turbot; and the young condition of the last-named fish, when about the size of Donovan's figure, is marked with dots on the under side as in this plate.

The Brill, or Kite, is often met with on the north coasts of Europe, and from thence along the borders of the British Islands to the Mediterranean; in which last district it appears to have been often confounded with the Turbot. But they seem to be in less numbers than the last-named fish, or at least they are less frequently caught with the line, which circumstance tends to shew that these fishes are less ravenous in their appetite than the Turbot. Their habits lead them to keep in similar ground, and their food is much the same; but the Brill is not considered as equal for the table to that esteemed delicacy.

The Brill reaches the ordinary size of the Turbot, but never is found of the bulk of the larger examples of that fish. Its form is more oval, and consequently is somewhat longer in proportion to its breadth: the dimensions being that the breadth of the body is contained one and two thirds in the length, excluding the caudal fin, and the length of the head three times and a third. It may be also distinguished from this only other British species, with which it might be confounded by having a slight covering of scales on the coloured surface; but especially by not having bony tubercles, such as are scattered over the skin of the Turbot. The gape is large, the angle of the mouth depressed, mystache rather wide, reaching fully hack to the lowermost eye; under jaw protruded, with a chin; teeth in the jaws and palate. The lowermost eye in advance of the upper; hindmost gill-cover passing a little over the root of the pectoral fin. Body and cheeks covered with scales; lateral line arched over the pectoral. The dorsal fin begins in front of the eyes, and but little above the snout, the rays stout, fleshy, and at first passing beyond the membrane, with from seventy to eighty rays; pectoral nearly round; ventral near the throat, with five rays, but this fin on the lower side passes behind the upper, so as to appear joined to the anal; the latter ends opposite the dorsal, near the tail; this last-named fin round.

The general colour of this fish is deep brown, mottled with deeper brown, and irregularly dotted over with white specks; and I have seen an example intensely black, with a few white specks on the anal fin. Like others of this family, the lower side is sometimes coloured like the upper, and in some instances only a portion is so coloured.

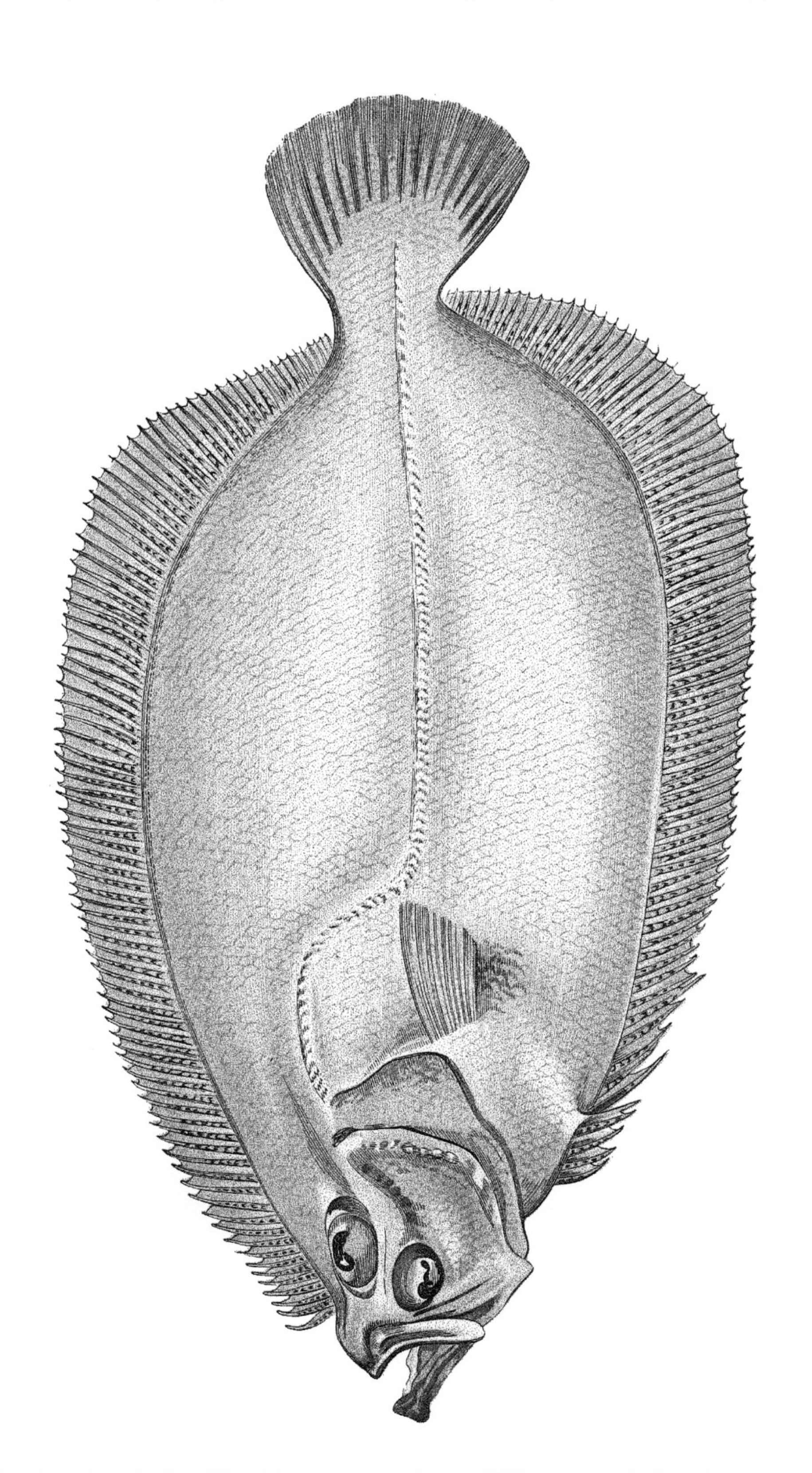

SAIL FLUKE.

Zeugopterus velivolans, RICHARDSON; 2nd. Supplement to Yarrell's British Fishes.

THE history of this fish, as it is given to us from the Islands of Orkney, is so extraordinary that doubt might be raised of its truth, if it were not supported by the evidence of scientific observers, who have used their endeavours to place the particulars beyond reasonable question. The earliest account appears to have been furnished by Dr. Duguid, of Kirkwall, but it is since confirmed by information received from Mr. Charles Thomson, of North Ronaldshay, who adds that this fish is scarcely rare in that distant island; and in others of that group it may be scarce only from its peculiar habits, for it has not been known to take a bait, and it is only in a single instance that it is said to have been caught in a net. The chance therefore of its falling into human hands was not to be calculated on, and still less so, as we shall see, in an unmutilated condition. But its characteristic habit is described as, that in the winter, when the weather happens to be calm, it rises to the surface of the sea, and with its tail lifted as a sail it makes towards the land, on the sandy shore of which it drives itself, and where it does not fail to be descried by some one of the larger prowling gulls; which, always on the look out, pounces on it for the purpose of tearing open its bowels in search of the liver; and when it has whetted its appetite with this delicacy it bears off the body of the luckless fish to some safe place where it may be devoured at leisure. As the eyes of these birds are sharp to discover this coveted prey, it is not often that the fish, which is said to be delicious when quite fresh from the sea, can be procured for the table; but to obtain it at the favourable moment of time boys are set on the watch, and it sometimes happens that by them the gulls are disappointed of their prize.

The interest attached to the possession and examination of such a fish has necessarily led to enquiry after it in the regions to which it was supposed to be confined; and in doing this I have been materially and kindly aided by the proffered assistance of Mr. John G. Iverach, of Kirkwall, whose earlier efforts, however, were only successful in procuring the information that the Sail Fluke had not been met with there for several years. At last, however, we have been more fortunate, and two examples were obtained in the Island of North Ronaldshay, by Mr. Charles Thomson; who has kindly contributed to the progress of scientific knowledge by sending them to Kirkwall to the care of Mr. Iverach, by whom they were dispatched, enclosed in salt, to me; and it is from these specimens that our figure and description have been taken. I have not been informed of the manner in which these examples were caught; but it is proper to observe that in this instance the gulls have been deprived of their banquet, for no mutilation was to be discerned, and the liver remained within its proper cavity.

From examination, however, we are led to conclude that, although the remarkable habits above referred to as being observed in Orkney, have not been noticed in our south and western districts, the fish itself scarcely appears to be rare with us. It is judged by Dr. Gunther to be the Whiff, (*R. megastoma* of authors;) and although we find it to differ in some portion of its character, as we shall see, from that of the subgenus in which that gentleman arranges it, it so closely approaches to what we know of that species, that we do not question the fact; although from comparison, as they lie together, we are further led to conclude that the Carter of Cornish fishermen must be set down as a distinct species. The difference between them, as well regarding structure as habit, will more particularly appear when we have described the last mentioned fish; but at present we only remark that the English examples of what we suppose to be the Sail Fluke, like those of Orkney, have not been known to take a bait; nor has it been found enclosed in nets that have been used near the land. It is in the trawl only that we have known them obtained at Plymouth and Falmouth: and it is probable that where the same method of fishing is employed, it is not rare along the south and eastward coasts of England and Ireland. Mr. Thompson found small fishes in the stomach of Whiffs caught on the east coast of the last named country.

The Sail Fluke grows to the length of about twenty-three or four inches, but that one which we select for description measured only seventeen inches and a half. The other example sent with it exceeded this a little in length, but in form and proportions they were otherwise closely alike. The greatest breadth or depth of the body was seven inches, exclusive of the fins, the body plump and thick. The head from the snout to the border of the gill-covers a little more than a third of the length of the body to the insertion of the caudal rays; the breadth of the body is equal to the length from the snout to the angle of the curve of the lateral line. Gape of the mouth wide, opening obliquely downward; under jaw protruding beyond the upper, with something like a chin; teeth in both jaws; the front end of the vomer bent down into the form of a protuberance, which is armed with teeth of which the points are directed backward. The tongue free, narrow, firm, and pointed. Eyes large, oval, the lowest smaller than the other and more advanced, coming near to the mystache, which is long, reaching to the middle of the lower eye; a ridge between the eyes, curved upward posteriorly. Above the point or symphysis of the upper jaw is a prominence, close behind which there is a depression from which the outline rises again to the back. Lateral line arched, rising at first and then sinking behind the border of the pectoral fin, from whence it makes a sharp turn towards the tail. Scales on the gill-covers, head, and body, more prominent on the lateral line, their edges finely ciliated, the largest towards the tail. The dorsal fin begins close behind the prominence on the snout, and, as does the anal becomes widest near its termination near the tail the root of the tail spreading wider as it proceeds from the ending of the dorsal and anal fins; first rays of the dorsal slender, lengthened their extremities free. the whole number ninety-two In one example the ventral fins had five rays, in the other six, near the throat, rising on the body posteriorly, and in the space left between them rose the first rays of the anal fin in front of which was the vent.

opening into a hollow between the ventral fins. In the anal fin seventy-two rays, in the pectoral twelve, the first very short; on the under side this fin is scarcely half as large as that which is above. Rays of the dorsal and anal fins bent in a curve backward towards their extremities. In the caudal fin eighteen rays, of which the middlemost are much the longest. The colour had probably faded through the length of time these examples had been out of water; the one described was pale brown without spots, in the other was a tinge of pink. Along the rays of the dorsal and anal fins was a series of slight scales.

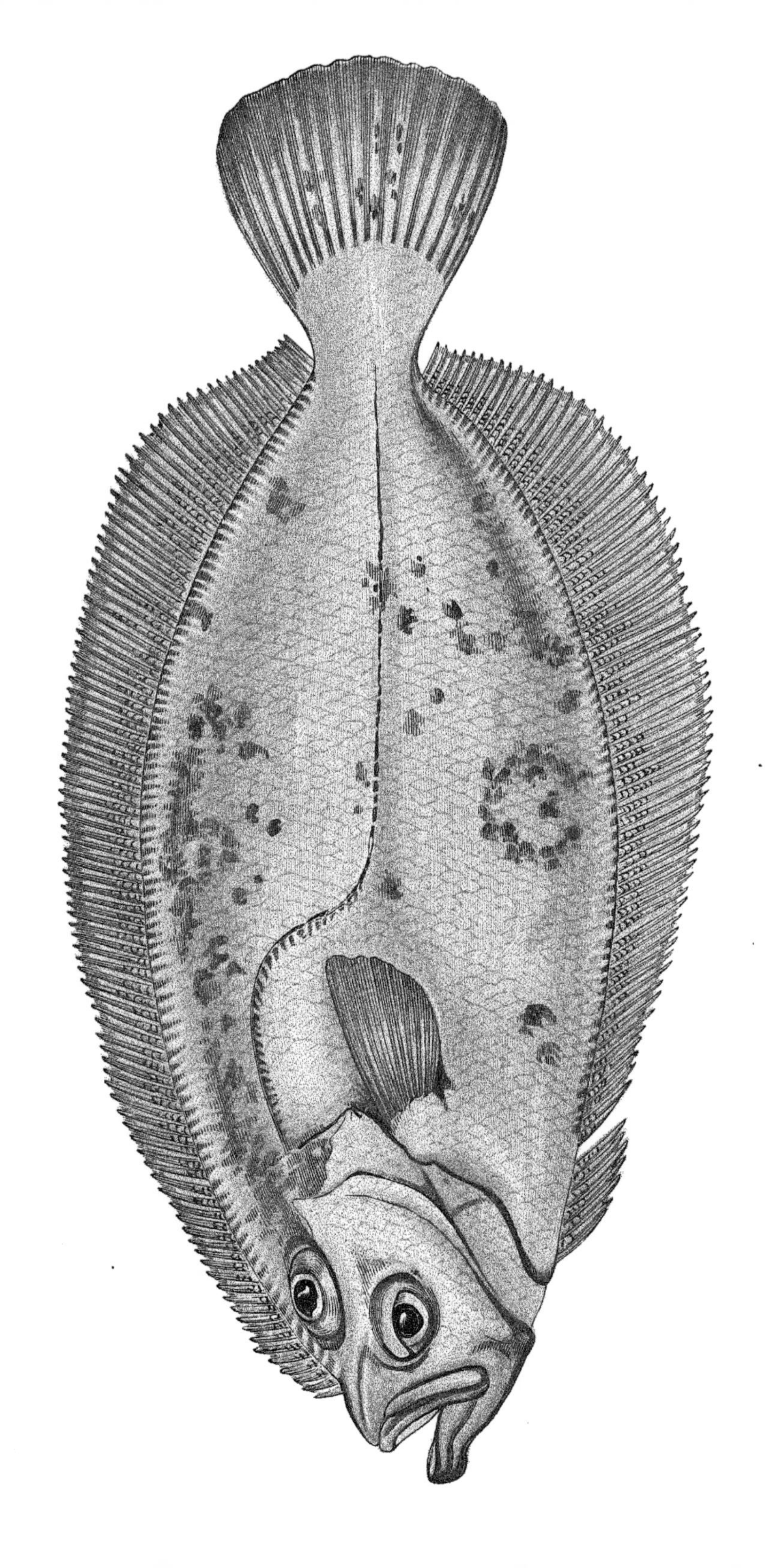

CARTER.

MARY SOLE. QUEEN'S SOLE.?

Whiff,	JAGO; Synopsis Piscium.
"	RAY; pl. 2, f. 2.
"	LINNEAN TRANSACTIONS, vol. xiv, p. 78.
Pleuronectes megastoma,	DONOVAN; pl. 51.
Rhombus megastoma,	YARRELL.

THIS may be called a common fish along the south coasts of Britain, but how far it answers to some of the flatfishes found in the Mediterranean Sea or south of Europe, which are designated by the appellation of *Oculata*, from the eye-like or circular spots which lie along the borders of their coloured surface, I have not the means of knowing, since the figures of the latter which I have seen are very imperfect, and the descriptions little less so. But in Cornwall the Carter is not unfrequently taken with a line, as well as in the trawl, and I have found a fish three or four inches long in its stomach; but it is little valued in the market on account of its meagre appearance, the quantity of flesh which clothes its skeleton being in less abundance than is common with any other of the British flatfishes. The body is indeed so thin as to authorise the name of Lanthorn Fish, by which it is sometimes called.

But there is an interest attending it arising from its having been confounded with the western examples of the Sail Fluke, under the common name of *Rhombus megastoma;* from the fact that both of them are alike characterized by a large gape, which is indeed the chief particular that marks this section of the family of flatfishes. But a glance at the figures of each of these fishes will shew that in their relative proportions they differ much. When full grown the Sail Fluke reaches a larger size, one example measuring twenty-one inches, and the weight of another was four pounds, and in plumpness it much excels the Carter. The comparative breadth of the former is also more considerable, being equal to the length from the snout to the angle of the curve of the lateral line; while in the Carter the same breadth (or depth) is only equal to the length from the snout to the extent of two thirds of the pectoral fin, the breadth of the body being also carried backward more considerably towards the tail. In the Sail Fluke the measure of the head, from the snout to the extreme border of the gill-cover, is a little more than a third of the length of the body as it ends in the insertion of the rays of the tail, which exceeds a like measure in the Carter, in which fish the middle rays of the tail are less extended.

But the difference between these fishes is particularly to be noticed in the difference of situation of their ventral fins, the one or two first rays of the anal fin in the Sail Fluke being embraced within the ventrals, which also conceal the vent; whereas in a Carter of the same length the space between the last rays of the ventrals and the first of the anal is about half an inch. And this circumstance, with the measurement in the Sail Fluke, is the more worthy of notice, as both of my specimens were alike in this respect, while the figure engraved in Sir John Richardson's (third) "Supplement to Yarrell's British Fishes," represents a space between the ventral and anal fins.

For the better distinguishing between these fishes we add a particular description of the Carter:—The extreme length of the specimen was eighteen inches; the depth, about the middle of the body, and exclusive of the fins, six inches; from the point of the lower jaw to the hindmost border of the gill-cover four inches and a quarter. Gape wide; beneath the lower jaw a firm, pointed, angular process. Several rows of conical, somewhat incurved teeth, with a vacancy among them at the symphysis of the jaw; a very small patch of teeth in the palate, with a pair that are larger than those in the jaws. Body thin; head bony; eyes large, the lowermost moderately prominent, the upper sunk within a wide socket. Body and head, with upper part of dorsal and anal fins, covered with scales, which are ciliated at the edges; anterior nostril having a wide foliated border. Lateral line arched to the place where it comes near the pectoral fin, and from thence straight. The dorsal fin begins opposite the margin of the anterior eye, the rays bordered with a wing, and the connecting membrane deeply cut in; both it and the anal fin pass far back to the root of the tail, and then pass under it. Colour dusky yellow, in some examples deeper or reddish, with numerous irregular spots on the head, body, and fins, (mostly brown, but some light,) and round the borders of the body a series of large spots irregularly ocellated; sclerotic portion of the eye brown, and spotted, the margin bright yellow; under side white, tinged with red. The pectoral fin below smallest. Dorsal fin with ninety rays; anal seventy-two, bifid near their ends; upper pectoral fourteen, below twelve; ventrals above seven, below six; caudal seventeen, the middlemost longest. The tongue is almost conical, pointed, free.

MULLER'S TOPKNOT.

Pleuronectes punctatus,	BLOCH; f. 189. TURTON'S Linnæus.
" "	FLEMING; Br. Animals, p. 196.
Rhombus punctatus,	CUVIER.
Pleuronectes hirtus,	JENYNS; Manual. p. 462.
Rhombus hirtus,	YARRELL; Br. Fishes, vol. ii, p. 334.
Rhombus punctatus,	GUNTHER; Cat. Br. Museum, vol. iv, p. 413.

PENNANT appears to have been the first to notice this fish, but he seems to have had but a slight knowledge of it, as in the engraving he has given, in the first octavo edition of his British Zoology, he bestowed upon it the name of Smear Dab, which in his text he had already assigned to a very different species. But on the other hand it must not be imputed to him, but to his engraver, that the eyes in his figure are directed towards the right. Since the time of Pennant, however, a considerable amount of confusion has mingled itself with the accounts which naturalists have given of the characters of this fish as compared with another closely allied to it, which has sprung more especially from the belief that the distinction between them is to be recognised by marks which assuredly are neither constant nor decisive. A principal one of these is said to be that the one or two first rays of the dorsal fin in the species known as Bloch's Topknot, next to be described, are lengthened into a separate filament; but I entertain no doubt that this supposed mark is only of casual occurrence, and may as well be met with in one species as the other; as it is not uncommon also in the Turbot, Brill, already referred to, and, as we shall presently see, in the Megrim or Scaldfish. It is somewhat remarkable, however, that I have not seen this elongation of the first rays in any other species besides those of the genus *Rhombus*.

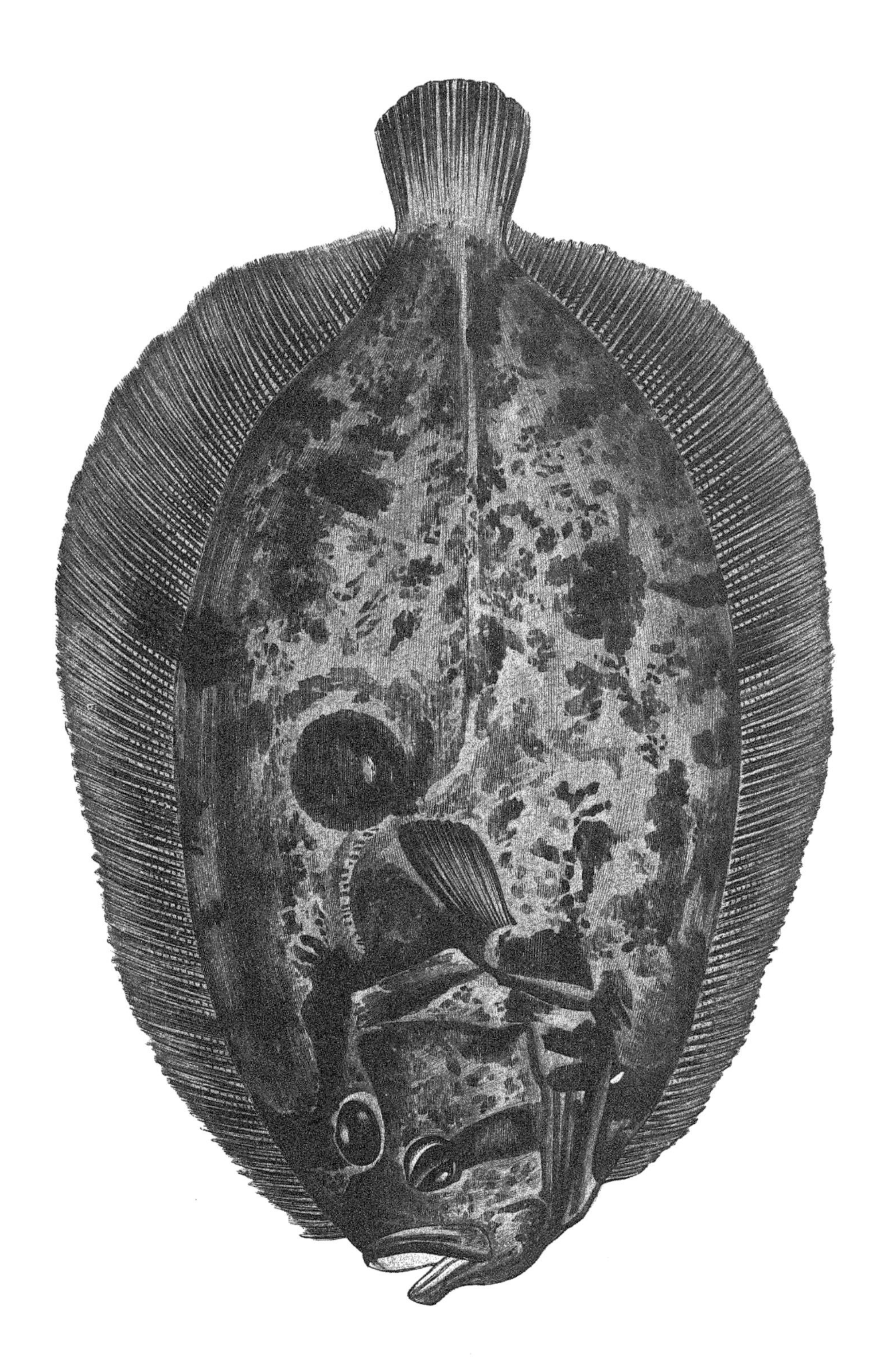

Another mark of distinction is described as existing in the roughness of the under side; but with respect to this we hesitate with the expression of a doubt as regards its constancy. There are variations of proportion in the parts of the body which may be observed when these fishes are brought together, but the separation of the ventral fins from the anal in Bloch's Topknot, whereas they are closely joined together in the present species, is a more decisive ground of distinction between them; although a question may arise whether the not very distant separation of these organs should be deemed sufficient to consign them, as has been done, to different genera. The underlapping of the termination of the dorsal and anal fins beneath the body, as here represented, appears to afford a

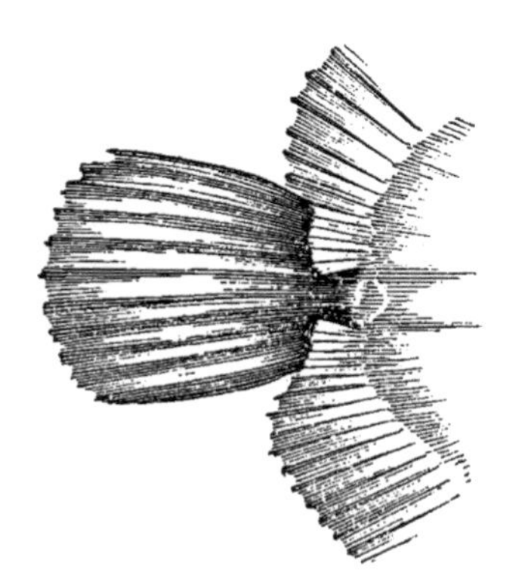

more characteristic distinction for constituting a separate genus, as distinguishing these two species of fish from all others of this family with which we are acquainted, as well in form as probably in habit.

The length of the example described which was of the usual size, was six inches and a half, and in width with the fins extended four inches and a half; but without measuring the fins the breadth was three inches; from the snout to the hindmost extent of the gill-covers one inch and six eighths; the gape wide, opening downward; the mystache wide, teeth small; the lower eye advanced before the upper. The lateral line rises in an arch over the pectoral fin, and from thence backward straight. The upper surface, as well of the head as body, rough from small scales, which also clothe the rays of the fins. The dorsal fin begins but little above the upper jaw; ventrals close under the throat, and united to the anal, and the latter with the dorsal broadest posteriorly; and then both of them becoming more narrow they pass under the body, so as almost to be brought together on the under surface. Pectoral and tail fins rather small. The colour is dark brown, with stripes and blotches that are almost black, especially on the head, where a broad stripe passes from each eye to their respective borders. There is often a broad circular spot, which sometimes is marked with a lighter centre, at the side where the arch of the lateral line ends.

This fish is subject to some remarkable variety. Thus, I have seen it with almost the whole of the upper surface, except the head and a small patch near the tail, without colour or scales, like the lower surface; the border of the dorsal and anal fins yellowish red, and with only two or three rays having scales; on the tail the rays extended beyond the border of the membrane. In another instance there was no caudal portion, so that where the dorsal and anal fins were brought closely together the body terminated.

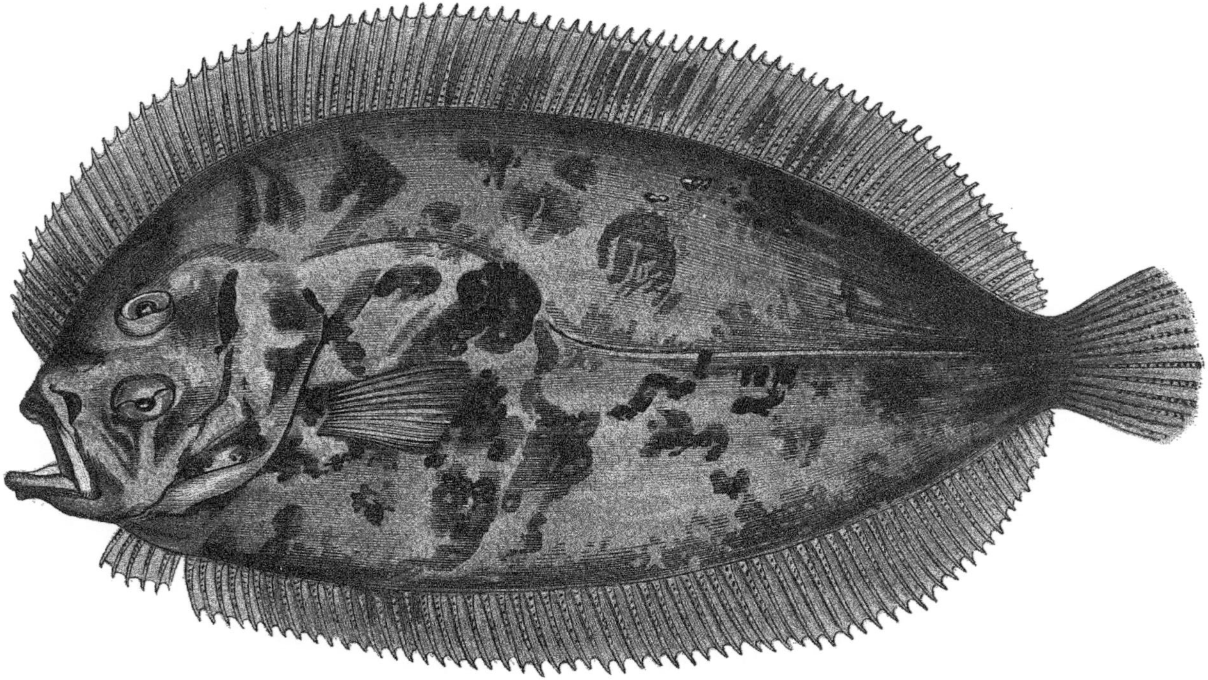

BLOCH'S TOPKNOT.

CLXVI

BLOCH'S TOPKNOT.

	BLOCH; pl. 189.
Rhombus punctatus,	YARRELL; Br. Fishes, vol. ii, p. 338.
" "	JENYNS; Manual, p. 462.
Phrynorhombus unimaculatus,	GUNTHER; Cat. Br. M., vol. iv.

THIS species has been confounded with Muller's Topknot, from which it requires some degree of discrimination to distinguish it; and even at this time a considerable amount of confusion exists as regards the assignment of their synonyms, as well as in the marks of distinction between them laid down by different writers. We have referred to these while speaking of the last-mentioned fish, and therefore it is only necessary in this place to specify those characters by which it may be definitely known, or which may seem to signify some difference of habits. In form they nearly resemble each other, but in Bloch's Topknot the width is scarcely so great; the gape is more limited, and the bony structure of the jaws and face less rigid; but the ridge between the eyes is a little more prominent, which, however, will scarcely be discerned unless the two fishes are laid side by side. The under side rough, but the only unvarying mark of distinction between them appears to be that in the present species the ventral fins are visibly separated from the anal, whereas in Muller's Topknot they are united. The colour also in Bloch's Topknot is generally of a lighter cast; but we cannot perceive sufficient reason for the trivial name given to it by Risso and adopted by Dr. Gunther, (*unimaculatus*, or One-spotted,) since both these species are often, and, it would appear, equally marked by a defined circular spot at the side, which, however, is sometimes not to be perceived. It is probable that the frequent lighter colour may be ascribed to its residence on lighter ground; and there is some reason to suppose also that Bloch's Topknot sometimes inhabits deeper water, as it has been taken from the stomach of a Ling which was caught at a depth of more than thirty fathoms. Mr. Thompson, of Weymouth, has informed me of an example that was ornamented on the lateral line, near the tail, with a black spot, which had an orange-coloured spot in its centre; and it was also marked with several puce-coloured spots along the base of the dorsal and anal fins.

In regard to the relative dimensions of this species Dr. Gunther says that the height of the body is nearly one half of the total length without the caudal, and the length of the head two parts out of seven; but in these proportions there appears to be an occasional variety, unless we are to suppose that a third species occurs on our coast. Mr. Thompson, of Belfast, obtained a specimen which, with a length of six inches and a half. was only two inches and seven lines in breadth, and I possess the note of an example which was brought to me that was six inches long, and two inches and a fourth in width. Another also, which was mutilated when I saw it, appeared to be of a length greater than the generality of these fishes. It seemed much flatter in form, the colour not dark, and the possessor, who had removed the lower surface before I saw it, assured me that the dorsal and anal fins ended in a line with the tail, without passing to the lower surface. The number of the fin rays was,—dorsal eighty-eight, anal sixty-eight, ventral six, pectoral ten, the caudal fifteen. It remains yet to be seen whether either of these was the *Rhombus cardina* of naturalists.

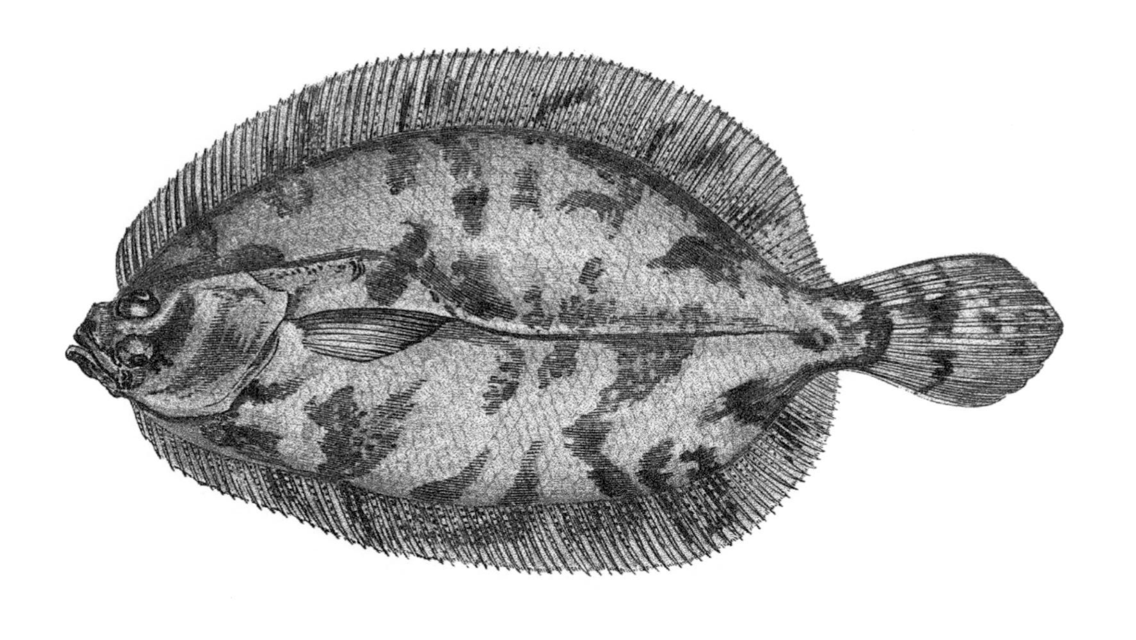

EKSTROM'S TOPKNOT.

CLXVII

EKSTROM'S TOPKNOT.

Rhombus cardina, FRIES AND EKSTROM: Scandinavian Fishes, pl. 50.
" *norvegicus*, GUNTHER; Cat. Br. M.. vol. iv, p. 412.

THERE is much confusion in the two or three writers I have been able to consult on the fish to which I have assigned the English name as above; and there cannot be a doubt that the name *Rhombus cardina* given to it by the Swedish authors, in their work on Scandinavian fishes is misapplied, as it bears little likeness to the species which is thus named by Cuvier, who refers to Jago's figure in Ray's Synopsis, f. 2, where it is represented under the name of Whiff. This last named fish, *Rhombus cardina* of Cuvier, is the Carter of our work. But on the other hand, Ekstrom's Topknot so nearly resembles the species of Topknots we have already described, that I have no doubt it has been hitherto confounded with them, although when seen together the difference is easily discerned.

The example from which our figure and description have been taken, is the first that has been recognised in Britain; and it was caught in the Bristol Channel early in the year 1863, from whence it came into the possession of Edmund T. Higgins, Esq., to whom I am indebted for the opportunity of making it known to naturalists.

Its habits appear to be little known even to the Swedish authors, Fries, Ekstrom, and Nilsson, who mention it; but there is reason to believe that its resort is less in rocky ground than the two other kinds of Topknots. Compared with them the proportion is not nearly so wide in comparison with the length, and it is also much thinner; gape wide; angle of the mouth depressed; lower jaw a little protruding, with a small chin. Eyes near each other, separated by a thin high ridge; the left eye in advance, and nearly touching the mystache; both eyes nearer the snout than in the *R. hirtus* of Yarrell, with which the comparison was particularly made. The gill-covers also much less marked with ridges: anterior portion of the head and jaws rough with points, but less so than in the kindred fish. On the cheeks and body the scales are regular and plainly visible, while they are not discerned in *R. hirtus*, and they are ciliated at the edge; but the whole surface, as well of the cheeks as the body, is smooth when the finger is passed over it to the tail. Lateral line less arched at first, by not ascending, and from the part where it is bent down proceeding straight to the tail. The dorsal fin does not begin so near the snout, but barely in front of the upper eye. The pectoral fin is longer and more pointed, for where in *R. hirtus* it barely touches the angle of the lateral line, in this fish it reaches beyond it; ventral fins not quite so close to the throat, but near the anal fin; tail much longer. The dorsal and anal fins pass only a little under the body at their termination, but this termination is not so near the caudal as in the other species. The under or uncoloured side, as well of the cheeks as body, is covered with ciliated scales, which is not the case with Muller's Topknot; and these scales pass far beyond the termination of the dorsal and anal fins, so as to cover the origin of the rays of the caudal fin. On the lower side the pectoral fin is small. Colour of the upper surface yellowish brown, mottled with darker brown over the head, body, and fins, with less tendency to defined spots than in the other species.

As two other species of Topknots have obtained their English names from eminent naturalists who have described them, it appears an act of justice to apply to the third species the name of a naturalist to whom we are indebted for a correct knowledge of the fish we have now added to the British catalogue.

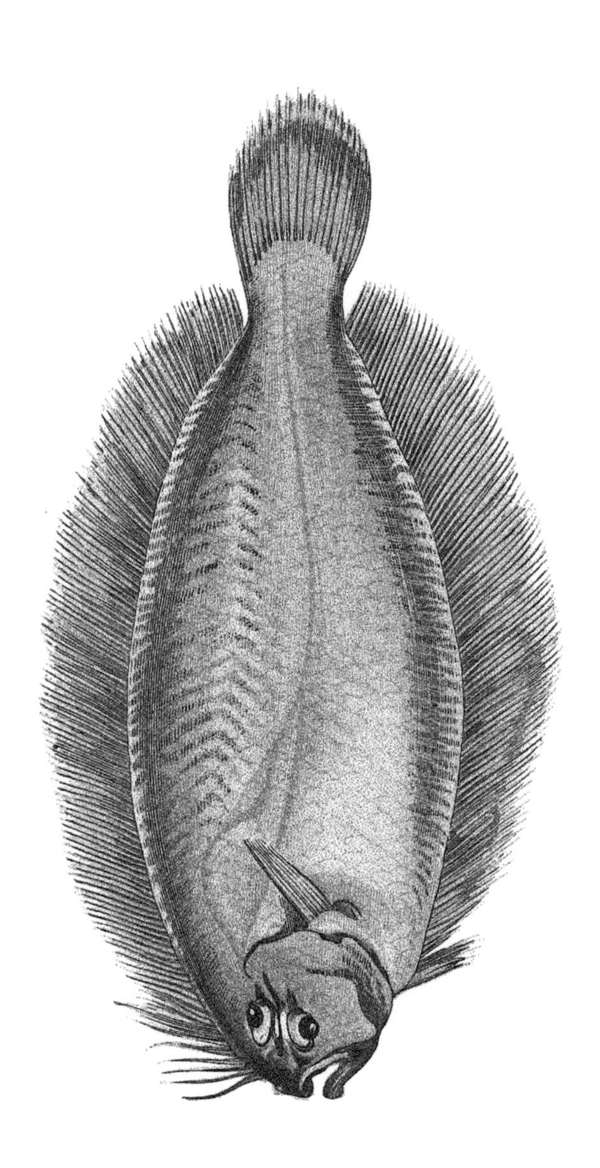

MEGRIM.

SCALDFISH.

Arnoglossus Rondeletii,	WILLOUGHBY; p. 102, Table F. 8, f. 7; copied from Rondeletius.
Pleuronectes arnoglossus,	FLEMING; Br. Animals, p. 197.
" *arnoglossus,*	JENYNS; Manual, p. 465.
Rhombus arnoglossus,	YARRELL; Br. Fishes, vol. ii, p. 345.
Arnoglossus laterna,	GUNTHER; Cat. Br. M., vol. iv, p. 417.

THIS little fish is with us one of the least regarded of the family of Flatfishes, chiefly on account of its small size, which, when of full growth, seldom reaches six or seven inches, but more frequently it is found of four or five inches in length. It keeps in the deeper water on sandy ground, and I have not known it take a bait, which may happen however because at the depth of water it frequents the hooks are larger than suit its mouth; but it is often swept up with the trawl, and sometimes it is found in the stomachs of the larger fishes, among which the Conger is a principal devourer.

But the most remarkable portion of the history of this fish is that from which it has obtained the name of Scaldfish, and in which it differs from all the flatfishes with which we are acquainted. Even when caught with the least amount of injury it is found to have lost the greater portion of its scales, and it only requires a slightly rougher handling for it to suffer the loss of its skin also, so that the surface usually appears as if the fish had been dipped in boiling water. It may be on this account, and from the small quantity of flesh that covers its bones, that with us it is not employed as food; but Rondeletius speaks of it as a delicacy for the table.

The example selected for description measured four inches

and three fourths in length, with the breadth of one inch and a half; eyes near each other, large, looking to the left, separated by an elevated ridge, the lowest a little in advance. Jaws equal, the gape wide, angle of the mouth depressed, with a stout mystache. Body thin, expanded just above the eyes, but widest at about the first third of the length backward. The scales large and thin, but so loosely adhering to the skin that they are easily separated from it, so that it is not common to find more than a few present. The skin also appears to be easily rubbed off from the flesh. Lateral line arched at first, not descending beyond the level of the uppermost eye, thence sloping until it proceeds straight to the tail. When the skin is removed, a line is seen which slopes gently downward, until it passes onward to the tail. The dorsal fin begins in front of the upper eye, and commonly is narrow at its origin, becoming wider at half its length, but in the example described several of the first rays were considerably lengthened into separate threads; pectoral fin narrow; ventrals close to the throat, and separate from the anal fin, the latter running parallel with the dorsal, and both ending short of the tail. Caudal fin rounded. The colour usually is pale dusky yellow, but in the present example a brighter yellow; border of the dorsal, anal, and caudal fins of a lighter colour.

LOPHOTES.

Arnoglossus lophotes, GUNTHER; Cat. Br Museum, vol. iv, p. 417.

DR. GUNTHER remarks, in his description of this species from examples formerly in possession of Mr. Yarrell, that this collection was entirely composed of specimens of British Fishes with a few species from the Mediterranean. The situations from which they were obtained had been noted by him only in a small proportion of the specimens, and unfortunately no record has been preserved by him of the history of the specimens of this species. Nearly all the Mediterranean specimens are prepared in a uniformly peculiar manner, different from that in which the British specimens in general, and the three specimens of a Lophotes especially, were preserved. It is not at all improbable that these three specimens are British, and to these observations of Dr. Gunther I will venture to add, that it is probable these same examples were examined by myself at Mr. Yarrell's house, at which time I made a note of its being that gentleman's opinion that they formed varieties or monstrosities of the Megrim or Scaldfish; but that they appeared to me to differ considerably from other examples of the last-named—especially in the lower jaw, which was more protruded; in the separate lengthened tendrils in front of the dorsal fin, and especially in the singular shape of

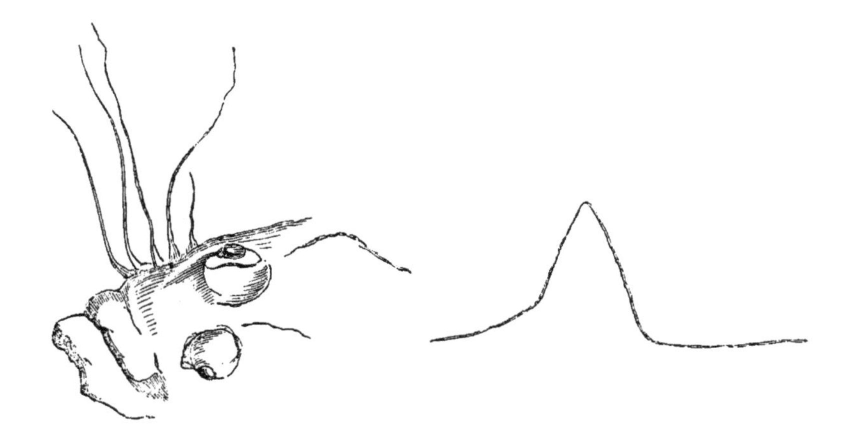

the lateral line; from which particulars I judged them to be a species new to Britain, but from whence they were procured did not appear. There is reason to believe that at least a large proportion of the preserved skins obtained by Mr. Yarrell from Plymouth were a gift from Lieutenant Spence, R.N., of that place; but I am informed by Dr. Gunther that these skins of the Lophotes in the collection of the British Museum are prepared in a different manner from such as were presented by Lieutenant Spence; and yet that this gentleman had met with the species at Plymouth appears from the fact that at his house I had an opportunity of examining a skin of what I felt no doubt was the same species with that of Mr. Yarrell, but of which I was only able to take a sketch of the front, with the form of the lateral line—a copy of which, a little diminished, is for the sake of illustration here given The skin now referred to was in length nine inches, and a very little more than three in breadth, which dimensions exceed those usually found in the more common Megrim; the head stouter and more bony than in the Megrim; eyes large, with an elevated ridge behind them: fin rays of the dorsal eighty-nine.

So far I am able to refer to my own notes of what has appeared to me to be an addition to the catalogue of British fishes; but in order to render what is known of the subject more complete, the following is added from Dr. Gunther, as above referred to:—"The height of the body is contained twice and two thirds to twice and four fifths in the total length (without caudal,) the length of the head four times and a half to four times and three fourths. Scales deciduous; anterior curve of the lateral line subsemicircular. Snout with the lower jaw slightly prominent, as long as the eye, the diameter of which is one fifth of the length of the head. The length of the maxillary is a little less than one third of that of the head. Eyes separated by a very narrow elevated ridge, the lower being in advance of the upper. The four anterior dorsal rays are elongate, nearly as long as the head. The dorsal fin commences in front of the upper eye, and terminates close by the caudal. Caudal fin somewhat shorter than the head, rounded. The length of the pectoral is two thirds of that of the head. Coloration uniform (in a dried state.") The dorsal rays were ninety-five, anal seventy-seven.

PLATESSA.

Mouth moderate, gape not large; cutting teeth in the jaws, none in the roof of the mouth. Dorsal fin beginning only so far forward as the upper eye, and neither dorsal nor anal coming near the tail. British species have their eyes towards the right.

PLAICE.

Plaice, Passer Bellonii, Quadratulus Rondeletii, Platessa Ausonii, Plaise,	Willoughby; p. 96, pl. F. 3.
Platessa platessa,	Cuvier.
" *vulgaris,*	Fleming; Br. Animals, p. 198.
" "	Jenyns; Manual, p. 454.
" "	Yarrell; Br. Fishes, vol. ii, p. 297.
Pleuronecte plie,	Lacepede. Risso.
Pleuronectes platessa,	Linnæus. Bloch, pl. 42.
" "	Donovan, pl. 6.
" "	Gunther; Cat. Br. Museum, vol. iv, p. 440.

Among the references which Willoughby makes to the names of this fish there is one to the poems of Ausonius, Epistle 4, where it is called *mollis platessa;* a designation which conveys the opinion held by some of our own day, that its flesh is too soft to form an acceptable food; while other writers have spoken of it in much more favourable terms. Nor is this difference of opinion to be ascribed altogether to variety of taste in those who have expressed it, since there is reason to believe that no inconsiderable variety is found in the quality of the fish itself, according to the situation in which it is caught; and this again is probably to be ascribed as well to the nature of the ground, whether it consist of mud or clean sand, as to the quality of the food on which it has been feeding; for the latter may well be supposed to exert an influence on the delicacy and firmness of its flesh. We have thus known, as an ordinary occurrence, that a great difference of price has existed between marine animals caught within a few miles of each other; and fishermen are quick to discern this in the appearance of fishes which are taken at different stations in the sea.

In the seas of Europe the Plaice is found considerably toward the north, so that it is known along the coasts of Sweden, and in the Baltic. It is also met with in the Mediterranean; but it is nowhere in greater plenty than in a moderate depth of water round the British Islands, where it forms an important object of the trawl fishery; and we are informed that the enormous number of thirty-eight millions of these fishes has formed the usual yearly supply to the markets of London. It is probable, however, that this amount is now greatly reduced, as I am informed by an old fisherman that so long since as the year 1833 the numbers taken by him on the coast of Cornwall had much fallen off from the time of his early experience. This large deficiency in his gains is ascribed to the destruction of the young and embryo fish, as well as of the fishing-ground and food, by the increased practice of trawling.

This fish, in common with several others of the same family, is taken with a line, and also with the spiller; the latter being, in fact, a representation on a small scale, as regards the hooks and snoozing, of the bultey or long line already described; the bait being the common worm of the beach, or shell-fish deprived of its shell, and the situation for the fishery some sheltered sandy bay where it may remain safe from interruption.

The Plaice sheds its spawn in spring; and Lacepede gives, on the authority of a friend, a curious account of the hatching of the young, which we copy, without being able to vouch for its accuracy or probability, as he also confesses for himself; but his final opinion is worthy of notice, as it may account for the conveyance of some kinds of fishes into situations where, under other circumstances, we could scarcely suppose they would be found. For a long time, says this author, the opinion has been held, as well in England as in France, that the Plaice is produced from a small crustacean animal of the shrimp kind; and it became the wish of Dr. Deslandes to find out how such an absurd opinion could have had an origin. He applied himself to the inquiry for several years, and, by way of experiment, he placed some of these little shrimps in a vessel of sea water of sufficient size; and in about twelve or thirteen days he found in the vessel eight or nine little Plaice, which gradually increased in size; and this circumstance confirmed him in the opinion, of the truth of which he had been already persuaded. In the following spring he pursued the inquiry by placing some Plaice in one vessel, and in another vessel some Plaice together with some of these small shrimps. It appeared that among the Plaice contained in these two vessels there were some females which shed their spawn; but, notwithstanding this, the only vessel which shewed the presence of young Plaice was that in which were the shrimps. Deslandes proceeded to examine the crustaceans, and then it was that he discovered the grains of the roe of the Plaice attached to the under part of these crabs. He opened these grains, and thus was able to ascertain that not only were they in a fertile condition, but that each one of them held an embryo which was somewhat advanced in development; from which circumstance he was led to form the conclusion that the grains of the roe of the Plaice can only come to life when hatched on the under surface of these small shrimps. But on the other hand, Lacepede was firmly persuaded that this opinion of Dr. Deslandes is founded on error, and that those grains of ova had been first shed in a place frequented by these shrimps, which are known to be disposed to feed on the ova of fishes, and especially on those of flatfishes. He further remarks that the ova of all kinds of fishes are enclosed in a tenacious kind of covering, and thus, as in this instance, they may become glued to the under portion of these shrimps at a time when the latter were seeking to devour them. Of course they would carry them thus attached wherever they themselves went.

I possess the record of a Plaice which measured nineteen inches in length, and ten inches in breadth; but these dimensions probably included the width of the fins. Mr. Thompson mentions an Irish specimen that weighed twelve pounds; and Ekstrom says he has seen it two feet long; but usually they are much less, the breadth of the body being about one half of the length, excluding the caudal fin. The most elevated eye is close to the upper margin of the body, which there suffers a depression in its outline; and the lower eye is in advance, near the end of the mystache, with a ridge between them, and six tubercles in a line backward, reaching to the lateral line. Lower jaw projecting; small teeth; gape rather small. Body and cheeks covered with small scales, which do not overlap each other; lateral line a little arched at its commencement and afterward straight. The dorsal fin begins with short rays close above the upper eye, and is widest behind the half of its length, ending at a short distance from the tail, as does the anal. Ventral fins about midway between the throat and the beginning of the anal fin; pectoral moderate; tail round. The colour varies in different examples from brown to deep bluish green; but both body and fins are dotted over with bright yellow or orange-coloured spots.

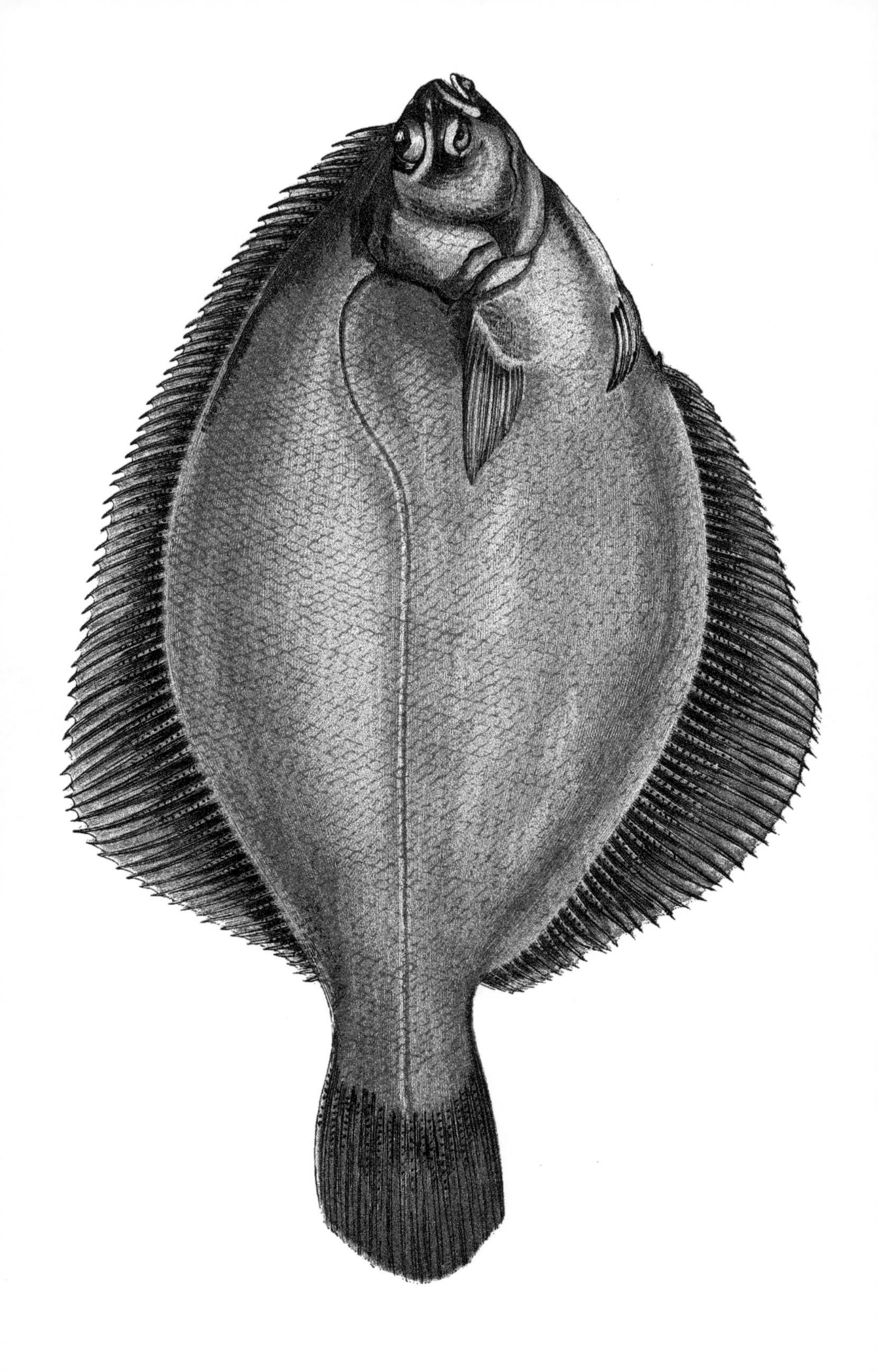

DAB.

Passer asper,	WILLOUGHBY; p. 97. pl. F. 4.
Pleuronectes limanda,	LINNÆUS. BLOCH, pl. 46.
" "	DONOVAN. pl. 44.
Pleuronecte limande,	LACEPEDE.
Platessa limanda,	CUVIER. FLEMING; Br Animals, p. 198.
" "	JENYNS; Manual. p. 456.
" "	YARRELL; Br. Fishes, vol. ii, p. 307.
Pleuronectes limanda,	GUNTHER; Cat. Br. M., vol. iv, p. 418.

THE Dab is one of our commonest fishes, but not one of the most abundant; nor does it appear to abound elsewhere, although it is met with in the far north of Europe. Its range, however, does not extend proportionally to the south, and although it is mentioned by Lacepede as an inhabitant of the Mediterranean, it is not named as seen in that sea by writers who have given an account of fishes which have occurred within their own observations. But it is well known along all the coasts of the British Islands, where it often takes the hook, and its food is worms, crustaceous animals, and small shell-fish. Its resort is in smooth and sandy ground, and frequently in sandy bays, although at times it is taken at the distance of several miles from land. I have found the milt ready to be shed at Christmas, but the usual time for spawning is in the spring. It is in esteem for the table as superior to the Plaice.

This fish rarely exceeds a foot in length, but the example described measured thirteen inches, with a breadth, exclusive of the fins, five inches and a fourth. The general form oval; gape rather small, under jaw a little protruding, with (in most cases) a small chin or tubercle; teeth stout, not close together; lowermost eye a small degree in advance; a ridge of moderate elevation between the eyes, in a recess of which is the largest nostril. Lateral line arched at first, and thence straight, running to the border of the tail. Body, cheeks, and the elevation between the eyes clothed with ciliated scales, which scarcely overlap each other. Dorsal fin begins opposite the superior eye, and with the anal ends short of the root of the tail; ventral fins separate from the throat and anal fin; tail round. The colour varies in this fish, the general tint being yellowish brown, often with clouds of deeper brown or yellow, and the pectoral fin is always yellow. I was favoured by the kindness of William Thompson, Esq., of Weymouth, with specimens in which, on a ground of yellowish brown, were numerous yellow spots, each one having a black dot within it; but these spots soon disappeared. It is to be remarked also that in these examples the middle rays of the tail were unusually elongated, and the dorsal and anal fins were bordered with white.

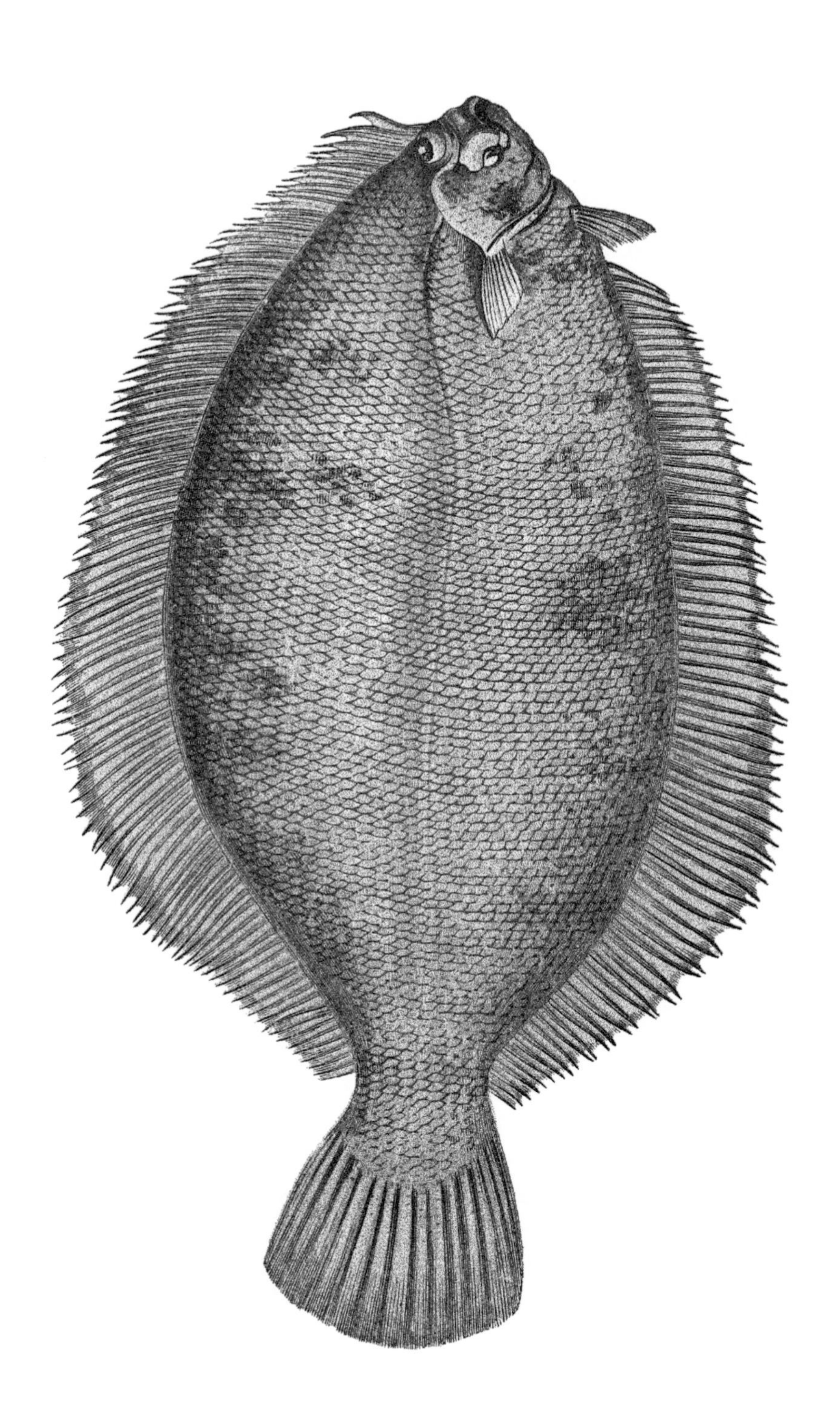

SMEAR DAB.

LEMON DAB. LEMON SOLE. QUEEN. SMOOTH DAB.

A Kitt,	JAGO; in Ray's Synopsis Piscium: but by this name Jago only meant to say that this is allied to the Brill, which fish is called Kitt, or Kite, in Cornwall.
Pleuronecte limandelle,	LACEPEDE.?
Platessa microcephalus,	FLEMING; British Animals, p. 198.
" "	JENYNS; Manual, p. 457.
" "	YARRELL; British Fishes, vol. ii, p. 309.
Pleuronectes microcephalus,	DONOVAN: pl. 42, but the head not well expressed.
"	GUNTHER; Cat. Br. Museum, vol. iv, p. 447.

THIS fish appears to have much of the habits of the Dab, and is equally common round the British Islands, except towards the more northern parts. I learn, however, from Mr. Iverach, of Kirkwall, that he has known it taken in Orkney in July; but it is less frequently caught with a hook: a circumstance little to be wondered at when we examine its mouth, which is so small that in the lesser examples it seems difficult to imagine how any but a minute object can be admitted into it. It is frequently caught with the trawl, and has a good repute for the table. The ground it frequents is for the most part stony; and it goes far to the north, as well as through the Baltic. It is prepared for spawning early in February, and appears to be among the most prolific of flatfishes; the lobes of roe being large, and extending back from the small abdominal cavity, compressed but wide, almost to the tail. The stomach and bowels are of slight texture, but the latter are large, and it is probable that the usual food is either vegetable or of the smaller sea insects, *(Entromostraca.)*

The Smear Dab in comparison with the Common Dab is a larger and thicker fish. The example selected measured in length seventeen inches, which is scarcely the utmost to which it sometimes attains; and, including the fins, the same fish was nine inches in breadth, but without the fins the breadth is two parts and a fourth of the whole length. The head small, the gape remarkably so, with tumid lips; teeth closely set, fiat, fifteen or sixteen in number, in regular order. Eyes moderately large, near each other, the lowest in advance and pressing on the corner of the mouth; anterior nasal orifice tubulous and projecting. Body and cheek clothed with smooth scales; lateral line slightly arched above the pectoral fin, and in one instance the arch interrupted with a depression. The dorsal fin begins over the upper eye and ends opposite the termination of the anal, not far from the tail, the latter round. Pectoral of rather moderate size; ventral midway between the anal and the throat. Colour reddish or yellowish brown with variegations of a deeper colour; under lip red.

In one instance an example, of which a plate is given, was caught with a line, the whole appearance of which was so different from that of the Smear Dab as commonly met with, that I have felt some doubt whether it should be assigned to that species; and my only reason for concluding it to be so is, that it still less resembled any other of the known fishes of this genus. The length was fourteen inches, and the breadth, including the fins, eight inches and a fourth; the head small, the distance from the lips to the borders of the gill-covers two inches and a half; lips tumid, gape small, teeth in an even row, with broad edges. Eyes large, protuberant, the lowermost in advance near the corner of the mouth, the two separated by a high ridge, and in front a high triangular space which comes over the snout, and is bent across the ridge to the other or lower side. The body remarkably thick anteriorly, humped on the nape; thinner towards the tail. Lateral line gently arched over the pectoral fins, and irregular as it approaches the tail; nostrils on the coloured side not sunk, the two pairs symmetrical—covered with minuter scales heaped together, slight scales on the fins and tail. The dorsal fin begins over the upper eye; first rays of the anal embraced by the ventral fins, which is not the case with the usual examples of the Smear

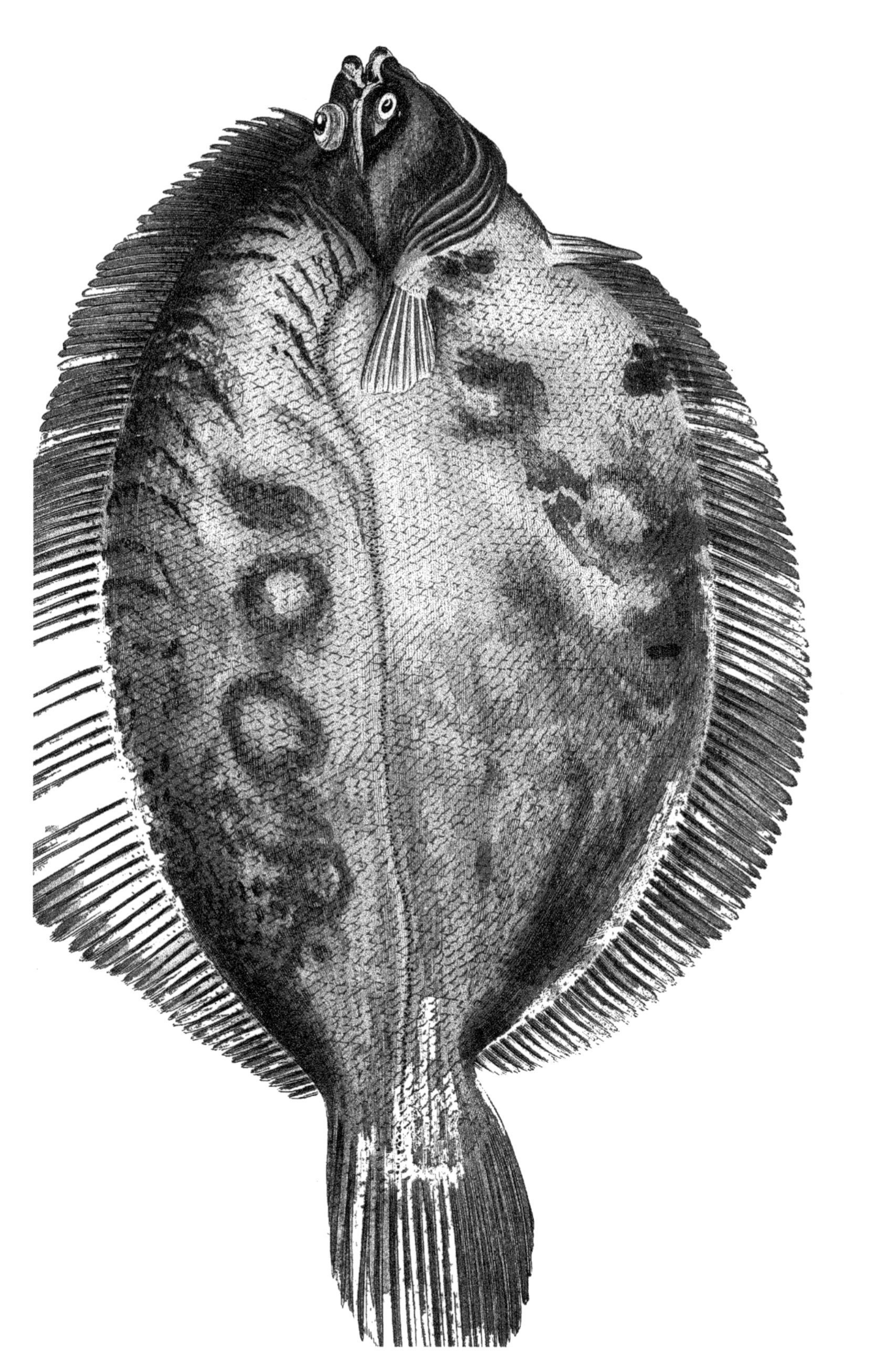

Dab; tail much more circular than in that fish. Colour generally a fine light brown, redder on the cheeks, the whole finely variegated, darker mottlings in irregular ocellated circles, bent lines along the roots of the dorsal rays, broken into spots posteriorly; fins more yellow. Fin rays—dorsal nineteen, anal seventy-three, pectoral nine, ventral five, caudal sixteen.

POLE.

CRAIGFLUKE.

Pleuronectes cynoglossus,	Linnæus.
Pleuronecte pole,	Lacepede.
Platessa pola,	Cuvier. Parnell; Memoirs of Wernerian Society, vol. vii.
" "	Yarrell; Br. Fishes, vol. ii, p. 315.
Pleuronectes cynoglossus,	Gunther; Cat. Br. Museum, vol. iv, p. 448.

The Pole is a fish of the Arctic Sea, in which direction it is found so far north as Greenland, where, according to Lacepede, in winter it keeps in the deeper situations which border on the shore; but this author also remarks that it is sometimes met with on the coasts of Belgium. In the British Islands it has been accounted amongst the rarest of flatfishes, so that when an example has fallen into the hands of a naturalist it has been thought worthy of special notice; and accordingly it is on record that some have been obtained on the coasts of Scotland, where this fish has been occasionally mistaken for the younger condition of the Holibut, with which, as well in some of its habits, and some degree of appearance, it possesses a general likeness. But there is reason to believe that it is to be ascribed rather to its places of resort, which are among rocks, or in stony ground where ordinary nets cannot be used, than to its absolute scarcity, that it is not more frequently taken with us, since it has been caught, at rare intervals, round the British coast, even in the West of Cornwall. But in its greatest abundance it appears to be local, as well perhaps, as migratory; and Mr. Thompson informs us that at times he has known it to be common on the east border of

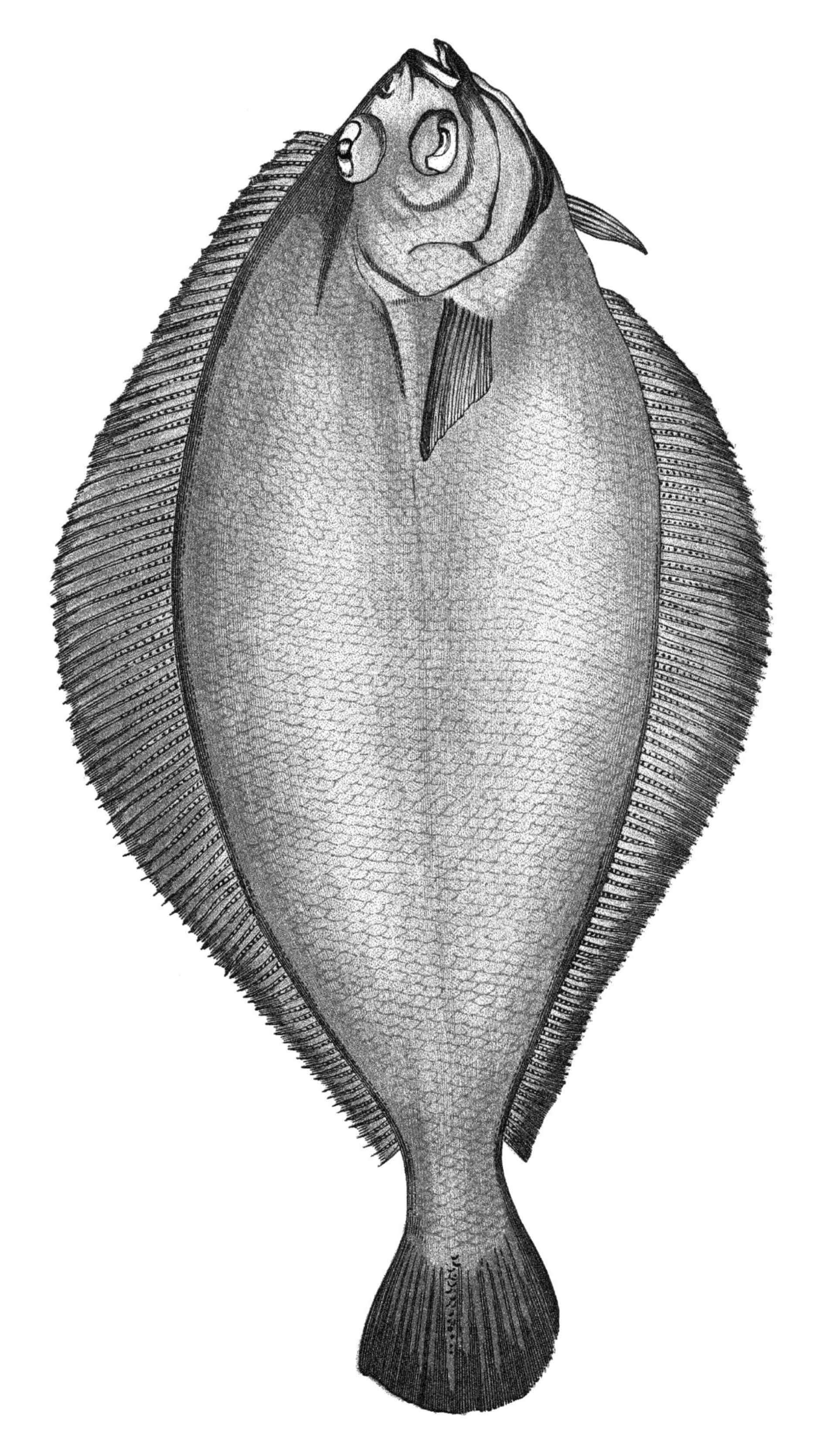

Ireland, so that one hundred and twenty of these fishes were in one instance caught at the single haul of a net, near Newcastle, in the County Down. Seventy also were obtained on another occasion near the same place; and it appears to have been so well known to the fishermen as to have acquired the name of White Sole, in distinction from what they term the Black, which is the Common Sole. But the former name is the less distinctive, as it is also applied to the Whiff or Carter, a fact which is so far descriptive of the state of Irish marine natural history, that there is scarcely a fish of that country which does not bear many names, even in places not distant from one another; and, more perplexing still, there are fishes not closely resembling each other which are so confounded together as to pass under the same denomination. It should be added, however, that in Scotland also this fish is not always known by the same name in every place. Bivalve shell-fish and crustacean animals were found in the stomachs of these fishes, and when offered for sale they found but little estimation in the market.

Dr. Parnell appears to have been the first who described the Pole from a British example, although at that time he was not able to assign to it any known synonym; and to secure greater accuracy we prefer to copy his description, taken from his paper on the fishes of the Firth of Forth, referred to at the head of this article.

The length of the specimen described was sixteen inches and a half, the breadth eight inches and a half, with a thickness of one inch; and in different individuals the proportions varied from twice and two thirds to three and a half of the breadth to the length, exclusive of the tail; the shape, therefore, is much like that of the Sole, but not quite so much lengthened. The gape small; jaws almost equal, or the lower a little the longest, with a row of blunt cutting teeth round each jaw; eyes separate, the lower eye a little in advance. Scales over the body and cheeks, but none before the eyes, of moderate size, with plain edges, and easily removed from their attachments. The lateral line at first descends slightly, afterwards straight. Ventral fins separate from the anal. The dorsal begins above the eye, widest at the middle, as is the anal opposite to it, the rays marked with fine scales; both these fins ending short of the tail, the latter fin with the middle rays longest. The colour yellowish or whitish brown, sometimes with spots on the fins. Dr. Parnell observes that it agrees with the genus *Platessa* in having the mouth entire, with a row of obtuse cutting teeth round each jaw, tail rounded at the end, with the eyes on the right side; but it differs from the Plaice in having no tubercles on the head, from the Flounder in not having a band of small spines on the side line, and from the Dab in not having the scales ciliated. From the Smear Dab it differs in having the lateral line nearly straight, the lower jaw longer, and the scales larger. It is to be noted that this is a different species from the *cynoglossus* of Rondeletius, which latter is called Pola in France, but which has the scales ciliated.

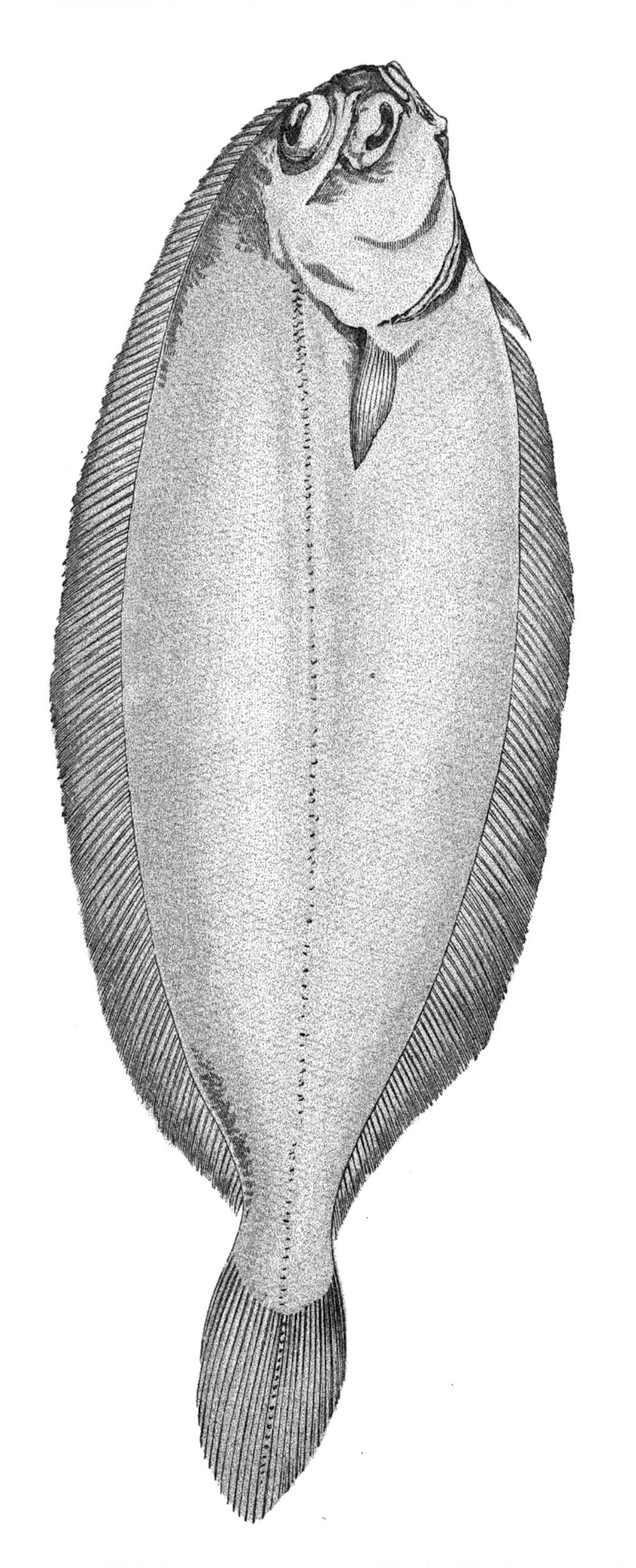

LONG FLOUNDER.

Platessa elongata, YARRELL; Br. Fishes, vol. ii, p. 318.
Pleuronectes elongatus, GUNTHER; Cat. Br. M., vol. iv, p. 451.

THIS is also regarded as amongst the rarer examples of British fishes; and so rare is it that hitherto two specimens only have been on record, one of which was furnished to Mr. Yarrell from Bridgewater Bay, in Devonshire, by Mr. Baker, at which place also the other was afterwards obtained. And even the example which was in Mr. Yarrell's possession has since been so far lost sight of, that it was not handed over to the British Museum when the collections of that gentleman came into the possession of that public institution. It is with much pleasure, therefore, that I find myself in possession of two examples, for which I am indebted to the kindness of Edmund T. Higgins, Esq., who procured them from Weston, in Somersetshire; and as this lies within the same district as Bridgewater, we may suppose that the range of this fish is exceedingly limited, and that, perhaps, even there its haunts are beyond the tracks of ordinary fishing. Our figure is from one of these examples, which, however, had lost its proper colour; but we choose to represent it in the condition in which we find it, rather than risk the danger of tinging it of a doubtful colour from description.

The example described was seven inches and three fourths in length, of which the body (exclusive of the tail) measured six inches and one eighth; breadth of the body one inch and six eighths; head short, measuring from the snout to the border of the gill-covers one inch and a fourth; under jaw protruded; gape moderately large; angle of the mouth depressed. Eyes large, the lowermost a little advanced, a prominent ridge between them. Body thin, shaped much like that of the Common Sole; scales rather large, but mostly denuded; lateral line straight. The dorsal fin begins close above the eye, with low rays; anal beginning below the root of the pectoral, and both end opposite each other, a short distance from the tail, and not much expanded in their course. Pectoral fin narrow, the one below smallest. Ventrals small, about midway between the throat and anal fin; the tail broadly lancet-shaped. On the cheeks of the blind side the scales are curiously arranged in separate small ovals. The dorsal fin has about one hundred and twenty rays, anal ninety-four, caudal twenty-six.

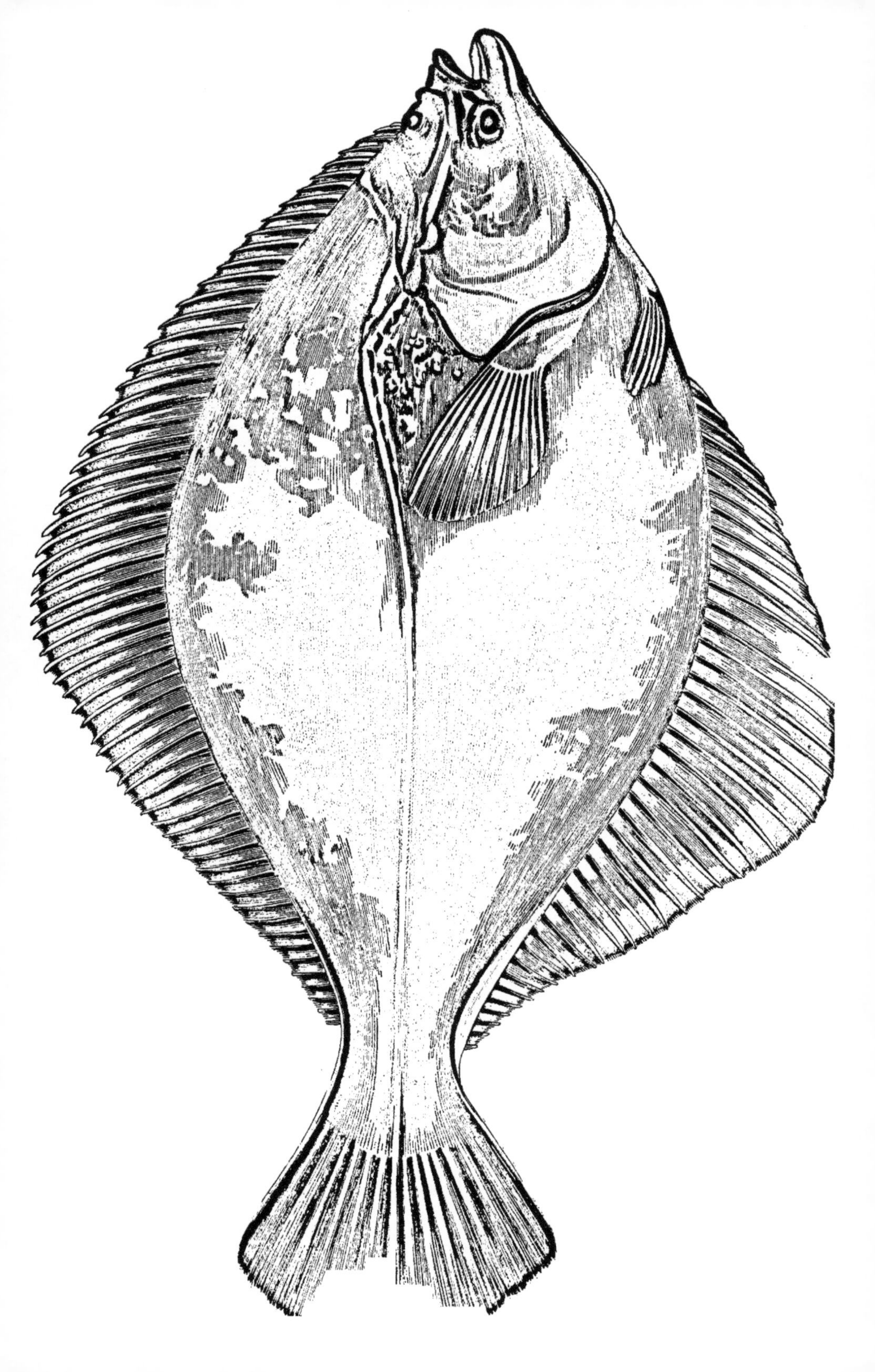

FLOUNDER.

FLUKE.

Passer fluviatilis, vulgo flesus,	WILLOUGHBY; p. 98, table F. 5.
Pleuronectes flesus,	LINNÆUS. DONOVAN; pl. 94, a variety.
Platessa flesus,	CUVIER.
" "	FLEMING; Br. Animals. p. 198.
" "	JENYNS; Manual, p. 455.
" "	YARRELL; Br. Fishes, vol. ii, p. 303.
Pleuronecte flez,	LACEPEDE.
Pleuronectes flesus,	GUNTHER; Cat. Br. M., vol. iv, p. 450.

THE Flounder is, more than others of this family, a fish of the shore, from which it never goes far; and it gives a preference to harbours into which a river flows, and which it traverses with the tide in search of worms or crustaceous animals; but it does not always retire with the sea, and fresh water seems at times to have a particular charm for it, as it occasionally wanders upward in the deeper rivers to a considerable distance, and there it assumes a new appearance as regards colour, as well as that it is said to suffer loss in the quality of its flesh, but it seems doubtful whether it ever breeds in fresh water.

This fish is found in abundance on all the coasts of the United Kingdom, and it is known even in Greenland; but the larger numbers appear to be on the north of Europe, where, however, it must be an uncommon circumstance to find it in such vast multitudes as is represented in the account furnished from the coast of Denmark, in December, 1862; where, as we are told in the "Zoologist," (April, 1863,) not less than two millions and a half were believed to have been taken after stormy weather, and the capture was at that time still going on, five hundred men having been engaged for three weeks in securing them. It is also known to be common throughout the Baltic, and, Ekstrom says, up to the sixtieth degree of latitude.

In the younger stages of their existence these fishes are devoured in large numbers by predaceous fishes and the larger diving birds, and an instance of this has been already given in our account of the habits of the Doree; but the loon (diver,) cormorant, and shag are more voracious enemies, while their sharp bills enable these birds to grasp them beyond the chance of escape. But to gulp down so wide a prey is not found so easy as to seize it, and it is amusing to the spectator to see the contrivance adopted by these birds to succeed in the attempt. The fish is to be first pecked in such a manner as to break or dislocate the bones, which can only be effected after repeated and violent efforts. The sides are then rolled together, like a sheet of paper, and with the head foremost the whole is safely passed into the capacious gullet. We may add here that if the prey be a crab it is taken to the surface, and the bird makes successive peeks at the legs, which, when struck with violence, are thrown off in succession by an effort of the animal, and duly swallowed, and the naked body is swallowed last of all. A Launce or Shanny is held aloft by the tail, or across the mouth, and then thrown into the air, when some skill is shewn in catching the fish with the head in the right direction, and it passes easily into the stomach.

I have found the roe of full size from the middle of December, the spawn being deposited in the tide-way of rivers; and I have also known the young to be excluded about the end of April, when they may be seen in the stiller parts of pools, their structure easily seen through, and moving in all directions, either flat or on their edge.

With us this fish seldom exceeds the length of a foot, and the heaviest I have a memorandum of weighed a pound and a half; but it is said to have acquired a much larger size. There is reason to suppose that the females are larger than the males. An example eleven inches and a half long measured four inches and a half in breadth, exclusive of the fins. The general shape bears a resemblance to that of the Plaice; gape moderate; mouth twisted, arched; under jaw a little protruded; teeth slender. Eyes rather large the lower one slightly in advance, and near the corner of the mouth. Scales on the body scarcely perceptible, but sometimes rough on the lateral line; rough tubercles along the base of the dorsal and anal fins; lateral line a little curved at first, but becoming straight before the termination of the pectoral fin, a bony ridge passing from the eyes to join it behind the gill-covers; and several rough tubercles on the coloured side, just above the pectoral fin. The dorsal fin begins with short rays close behind the eyes, and both the dorsal and anal fins end with short rays opposite each other, not far from the tail,—both these fins

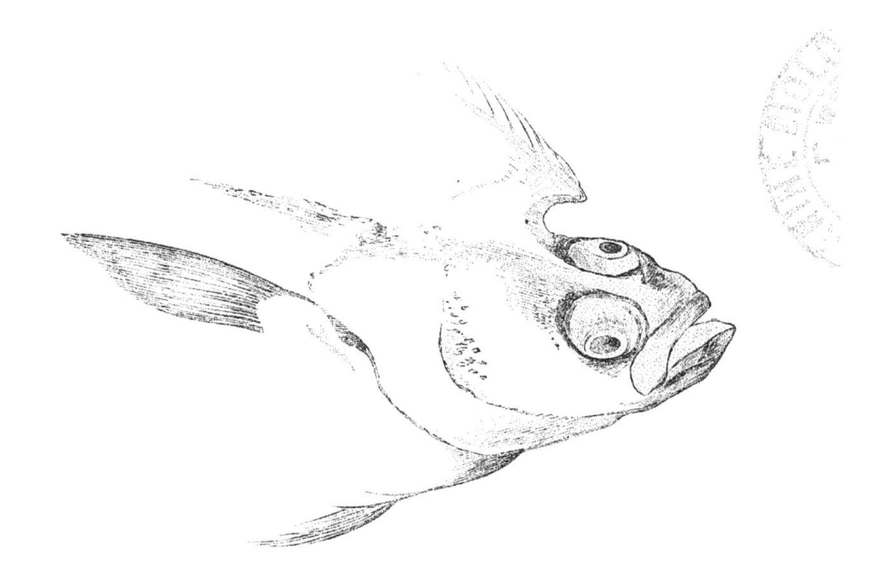

much expanded near the middle of their length; pectoral wide; ventral at midway between the throat and beginning of the anal fin; tail rounded. The colour is subject to variety, according to the nature of the ground, from very dark to a lighter greenish, or brown, with mottled tints, and the fins with even white streaks separating the rays.

It is said that this fish is more subject to variation of structure than others of this extensive family; the eyes being directed to the left instead of the right, or the lower and white side being of the same colour as the upper. Donovan's plate represents the opposite of this, with a large portion of the upper side like the paler side below. It sometimes happens also that there is a deficiency in the continuance of outline from the origin of the dorsal fin forward, and this has been accompanied with a black surface below; but it seems probable that this greater frequency of variation is only because the fish itself is more frequently caught than other species of the family of flatfishes. The yellowish spots which sometimes mark its sides are very different from those which adorn the surface of the Plaice.

But as regards the variety (here figured) which is marked with a notch or deficiency in the outline behind the eyes, and a black surface on the under side, it has occurred so frequently in some districts as to have raised the suspicion of its being truly a distinct species, the more especially as some difference in the fins has also been detected. But after comparing an example from the River Fowey with a specimen of Flounder of regular formation, I feel no doubt of their being specifically the same; while, on the other hand, I feel equally certain that this variety of structure in the outline is natural, and not, as has been supposed, the consequence of mutilation.

Some observations which have been made on the structure of the mouth and cheeks of this fish are probably applicable, with a little variation, to the other flatfishes already described; for the want of symmetry, which is the character of this family, extends to all the tissues or portions of the head, including the nerves of feeling as well as the muscles and bones, and thus creating in the upper surface a far higher degree of power and sensibility than exists in the lower. Thus, what appears to be the largest nerve in the body is seen to pass along the coloured side of the cheeks, to be distributed to the palate, in comparison with which there does not appear to be any one on the lower side, while there is another here situated which passes down to supply the angle of the jaw. On the coloured side the (masseter) muscle of the jaw is strong, and is united to that bone by a tendon, which is not the case on the white side; and a separate nerve, somewhat corresponding to that already mentioned on the white side, is given out from the skull proportionally further forward, from whence it descends under the tendon to the jaw, both these nerves being accompanied with blood-vessels.

SOLEA.

Form of the body oblong, the front rounded; eyes towards the right; mouth twisted toward the blind side, with teeth only towards that side. Lateral line straight. Pectoral fins on both sides of the body.

The fishes of this genus, like some others of this curiously-formed family, have received their name from the general appearance of their shape, as distinguished in their general outline. Thus the Common Sole is regarded as shewing the form of the lower portion or sole of a shoe; while with the Greeks it obtained the name of Bouglossa, from a supposed resemblance to the tongue of an ox. Hippoglossus and Cynoglossus were also named from a fancied likeness to the tongue of a horse and a dog. The Flounder is so called from its mode of progression along the ground, and its other name of Fluke is from its flat rhomboidal shape; being the same which is retained for the name, among other objects, of the flattened portion or claw of the anchor of a ship. *But* appears to be a northern word, which signifies a rounded and flattened surface, as was the mark at which arrows were aimed from a bow; and the word Tur, which was formerly written, and is still in many places pronounced Tar, added to But, is significant of a flatfish which has its surface studded over with thorns. It does not appear that the Holibut was so named from any idea of its supposed sanctity, but that it was best known near Heligoland, or the Holy Island. The name of Passer, by which some of these fishes were designated at an early date, was derived from a fancied comparison with the Sparrow; for no better reason than that both the fish and the bird were brown or dark above, and of a light colour below.

SOLE.

Buglossus, or Solea,	WILLOUGHBY; p. 100, table f.7.
Pleuronectes solea,	LINNÆUS. BLOCH; pl. 45.
" "	DONOVAN; pl. 62.
Pleuronecte sole,	LACEPEDE. RISSO.
Solea vulgaris,	FLEMING; Br. Animals, p. 196.
" "	JENYNS; Manual, p. 466.
" "	YARRELL; Br. Fishes, vol. ii, p. 347.
	GUNTHER; Cat. Br. Museum, vol. iv, p. 463, where it is classed in the section in which the pectoral fins of both sides are developed, and the nostrils of the blind side are not dilated.

THIS fish is one of the most common and abundant of the British flatfishes, as it also is among the most esteemed at luxurious tables. Its haunts are generally in sand or gravelly ground, and in deeper water than what is frequented by the Plaice and Flounder; although it is also said to come into the fresh water of tidal rivers, and even to thrive there. The range of the Sole in the ocean is very wide, since they are not only known round the British Island, but much further northward in the ocean, and through the Baltic, as well as through the extent of the Mediterranean, and southward even to the Cape of Good Hope; where, however, it is considered a rare fish. The large number of Soles which are caught in the United Kingdom may be judged from the fact that the average amount of those which were yearly brought to the London Market in the early part of the present century was ninety-seven millions, and those which were sold in other parts of the kingdom must have been proportionally great; but there seems little doubt that in consequence of such destruction the fishery at this time is much less productive. By far the greatest numbers are caught

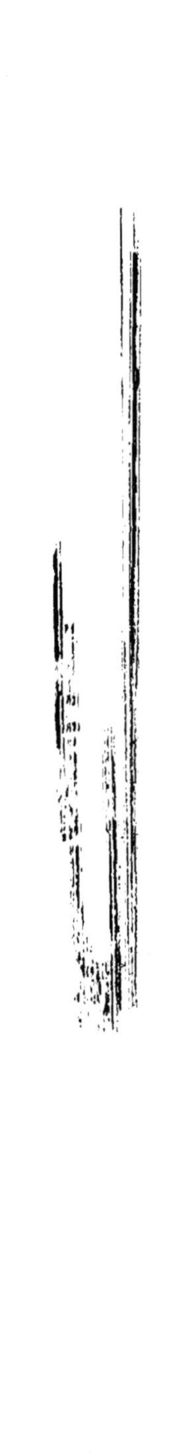

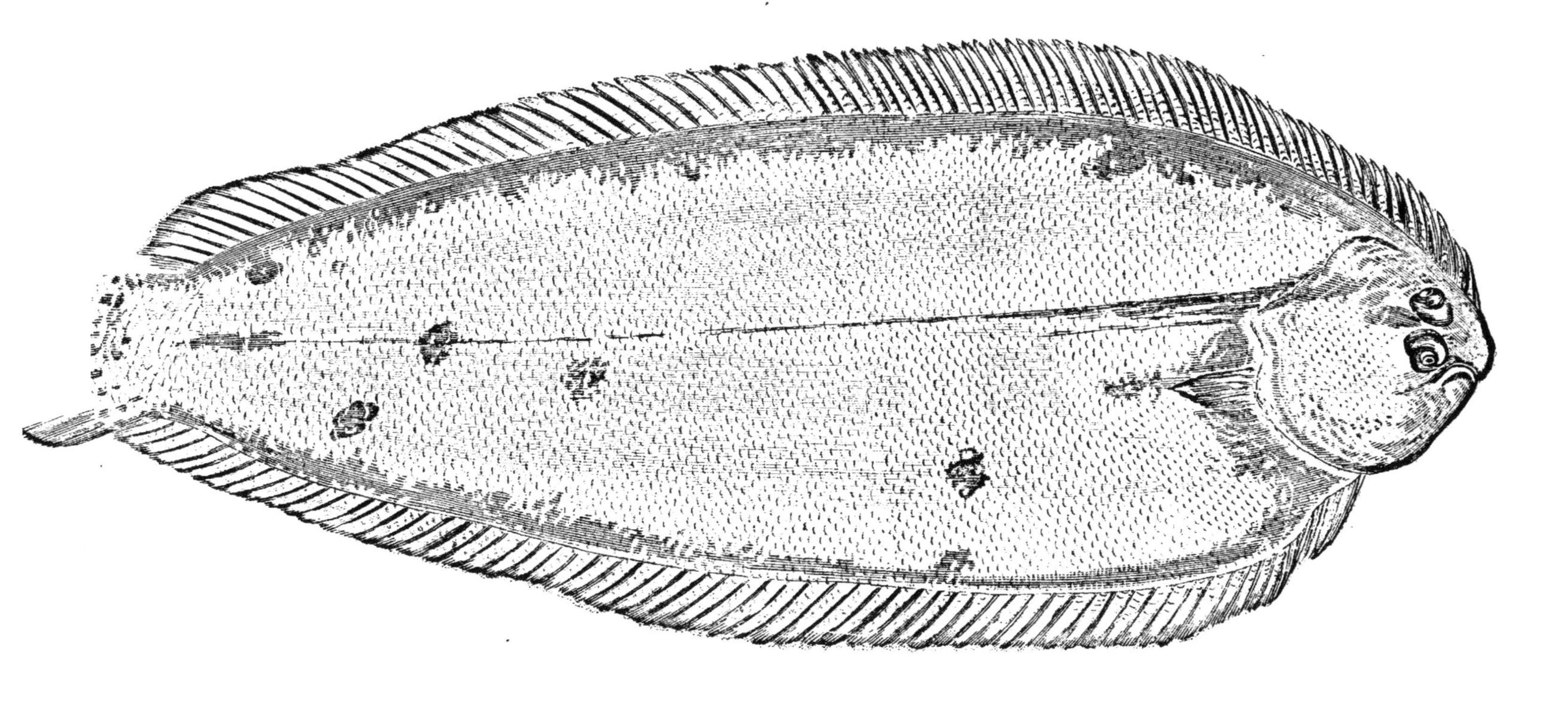

SOLE

with the trawl, from which, because of its form, this species has less powers of escape than most others of this family. Perhaps also this fish suffers less injury from this mode of capture, on account of the manner in which it moves; and yet we are informed that in the year 1832, there were condemned at Billingsgate, as unfit for food, twenty-four thousand six hundred Soles; with of Plaice eighteen thousand seven hundred and fifty; while of Turbots there were condemned only two hundred and seven, and of Brills one hundred and eighty. All these fishes are retentive of life, and do not soon decay after death; but the difference of numbers here referred to may arise from the circumstance that the two last-named species are more abundantly caught with long lines, where no violence is inflicted, while Soles are chiefly obtained by the trawl, where, when dragged for a long distance along the ground, the bruised bodies may speedily suffer decomposition. But the Sole is sometimes taken with a line, or rather on the hooks of a spiller; and that more are not thus caught is to be explained by the circumstance that this method of fishing is for the most part only followed by day, whereas the Sole usually seeks its food only by night; and by being aware of this, I have been informed by a fisherman that at one haul he once caught twenty-eight Soles, the bait being the lug or other worms of the beach. This fish spawns in the early portion of the year, and we may readily believe it to be prolific, since otherwise the immense numbers that are caught would soon extirpate the race.

Mr. Cocks mentions a couple of Soles, each of which measured twenty-three inches in length, with a breadth of ten inches, the weight nine pounds; but they are rarely allowed to reach so large a size. The example selected measured seventeen inches in length, and in breadth seven inches, including the fins; the general form flat; front of the head rounded, and protruding over the mouth, which is arched, the upper lip bent down over the lower, jaws twisted, with fine teeth on the lower or white side of the jaw. Eyes on a level, not touching the upper border or outline, the lower eye close to the corner of the mouth; a depression, and not a ridge, between the eyes; a very short barb in front of the eye. Head, body, and generally the fins covered with small scales; lateral line straight;

vent concealed between the ventral and anal fins, the latter overlapping each other. The dorsal fin runs from the arch of the head in front of the eyes, and anal from the ventrals, both to the root of the tail; the caudal fin round; pectorals small, their extremity blackish; colour of the head and body subject to some variation, but uniform dusky brown or greenish; under surface of the head spread over with woolly fibres; a prominent nostril above the lower arch of the mouth, and a smaller one far behind. Of the two pectoral fins the uppermost has four longer rays, the lower eight shorter rays.

VARIEGATED SOLE.

CLXXVII

VARIEGATED SOLE.

THICKBACK. BASTARD SOLE. RED-BACKED SOLE.

Pleuronectes variegatus,	DONOVAN; pl. 117.
" *Mangillii,*	RISSO.
Solea variegata,	FLEMING; Br. Animals, p. 197.
Monochirus variegatus,	YARRELL; Br. Fishes, vol. ii, p. 353.
Solea variegata,	GUNTHER; Cat. Br. Museum, vol. iv, p. 469.

THIS species is not mentioned among Scandinavian fishes by Nilsson nor Eckstrom; but its range extends to the north of Scotland, and it is known along the coasts of Ireland. Risso also met with it in the Mediterranean, where, however, he says it is rare. And yet this apparent scarcity may be rather from the mode of fishing usually practised than from the absence of the fish; as it was formerly said to be also little known in England, whereas in the markets of Plymouth and Penzance it appears equally plentiful with the Common Sole. I have not known an instance of its having taken a hook, even when the other Sole is sometimes caught; from which the conclusion may be drawn that not only is its food different, but that it also keeps in deeper water, where the smaller hooks are not employed. The smaller specimens are sometimes found in the stomachs of fishes from a depth of forty or fifty fathoms. It is ranked among the superior fishes for the table.

The Thickback seldom exceeds the length of eight or nine inches, and in its general proportions it much resembles the Common Sole, but with a little wider oval, and in its substance it is thicker. The head is rounded in front, where it advances beyond the mouth and lower jaw: the mouth itself small and twisted. The lower eye rather smaller than the upper, and in advance, close to the corner of the mouth. Lateral line straight, passing to the border of the tail. The dorsal fin begins near the upper jaw, and above the eyes is narrow; but afterwards it passes on in nearly equal breadth, and with the anal ends shorter of the tail than in the Common Sole; caudal fin more or less round; pectoral fins small, and on the under side especially so; ventrals also small, about midway between the throat and anal fin. The colour is reddish brown, and in the larger examples generally with few or no spots on the body; but there are several distant black patches on the dorsal and anal fins. In smaller individuals the dark mottlings pass across the body as well as over the fins.

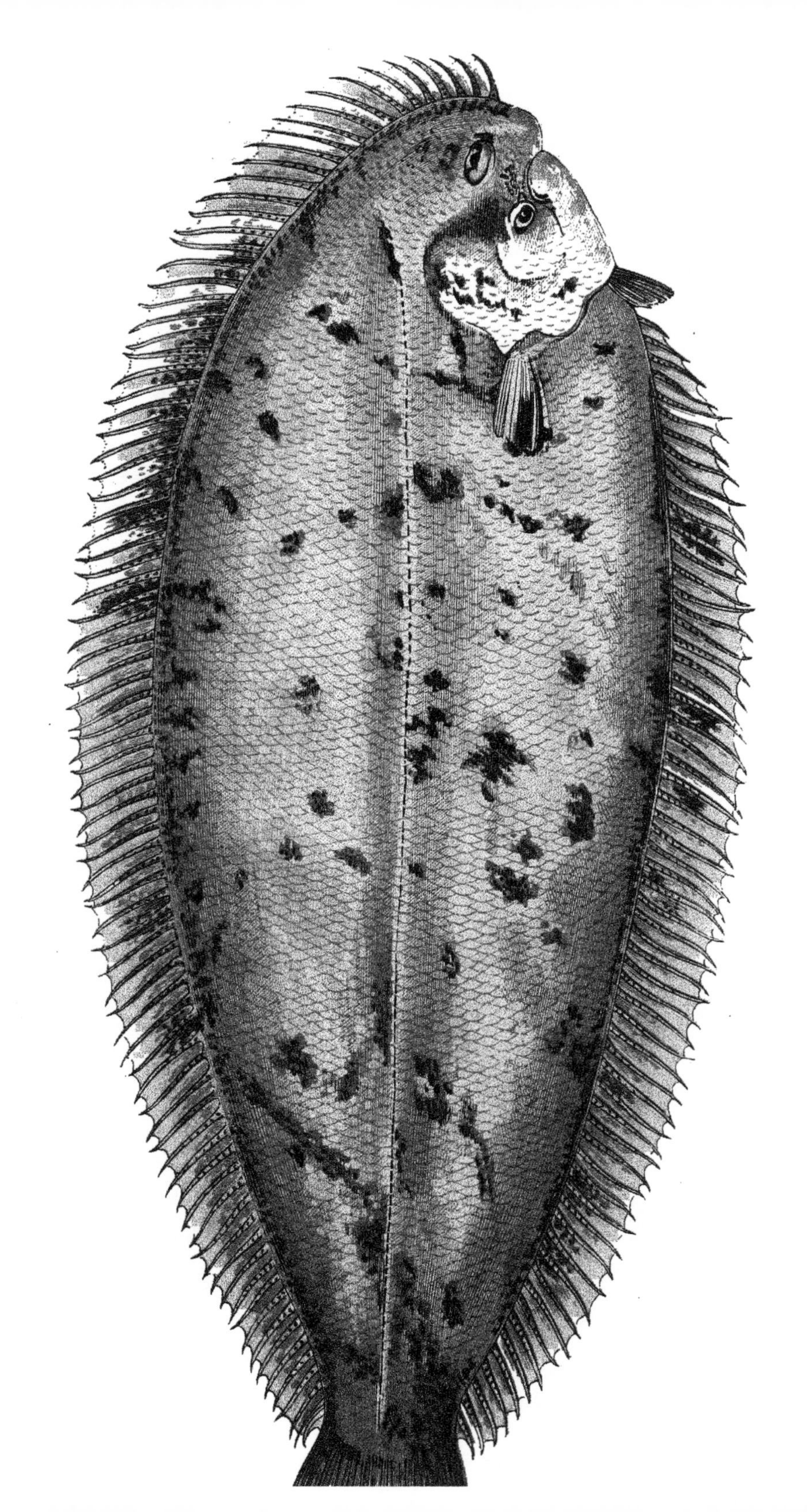

LEMON SOLE.

SAND SOLE.

Solea pegusa,	YARRELL; Zoological Journal, vol. iv, p. 467, pl, 16; and British Fishes, vol. i, p. 351.
" "	JENYNS; Manual, p. 467,
Solea aurantiaca,	GUNTHER; Cat. Br. Museum, vol. iv, p. 467.

THE Lemon Sole was first recognised as British by Mr. Yarrell, but was mistaken by him for *P. pegusa* of Risso, which is the *Solea monochir* of Dr. Gunther, and which is so called because it has no pectoral fin on the blind side; but that species has never been taken in Britain, although it is a native of the Mediterranean, where it appears that the Lemon Sole is also known. When first described by Mr. Yarrell, the latter was new to science.

It had probably been confounded with the Thick-backed Sole by fishermen in the south and west of England, to whom it appears to have been known to some extent. It seems, however, to be amongst the rarer of our flatfishes, although no small number have been taken in Ireland and in some parts of the east coasts of England; and its habits are no further known than that it prefers to keep in soft sandy ground. The single individual I have examined was taken at Plymouth, and was presented to me by Lieut. Spence, R.N., who occupied himself much in seeking out and preserving the skins of fishes taken in that port, many of which were sent by him to the British Museum.

The length of the specimen was eight inches and a fourth, which appears to be that to which this species commonly attains; breadth of the body two inches and six eighths. The forehead is round and projects much over the mouth, cousequently the under jaw is overlapped by the upper; mouth arched. Eyes separate, the upper distant from the border and advanced, the lower near the mouth. Scales cover the head and body, and also the lower surface, a row on each of the fin rays. Lateral line straight. The dorsal fin begins in front of the upper eye, and ends close to the root of the tail, being joined to it by a slight membrane, as is the anal; ventrals close to the throat; pectorals moderate. The colour of the fish yellow, studded over with spots, the pectoral fin having a black patch at the end. On the under side the nostril is broadly open. The number of the fin rays is—of the dorsal ninety, anal seventy-one, caudal eighteen, pectoral eight, ventral five. The difference of numbers in the dorsal and anal here given from those assigned to this fish by Mr. Yarrell, may be explained by the fact, (common to the flatfishes, as well as to all fishes which have numerous rays in those fins,) that, in addition to an intermediate bone attached to each neural spine of the vertebræ, and each one bearing a fin ray, several of them will have a couple of these intermediate bones, with each a fin ray; and as the number and distribution of these additional bones are not influenced by a regular law, the number of fin rays attached to them must vary accordingly.

By way of comparison with the Variegated Sole, when laid side by side with it, the scales appear much alike, but different from those of the Common Sole; while the more precise differences between the former fishes are—that in the Lemon Sole the eyes are smaller, the upper much more in advance, the mouth differently formed, the dorsal fin over the head more expanded, pectoral fin larger, ventrals nearer the cleft of the gills, dorsal and anal fins nearer the tail, and united to it by membrane. The colour and mottlings are very different.

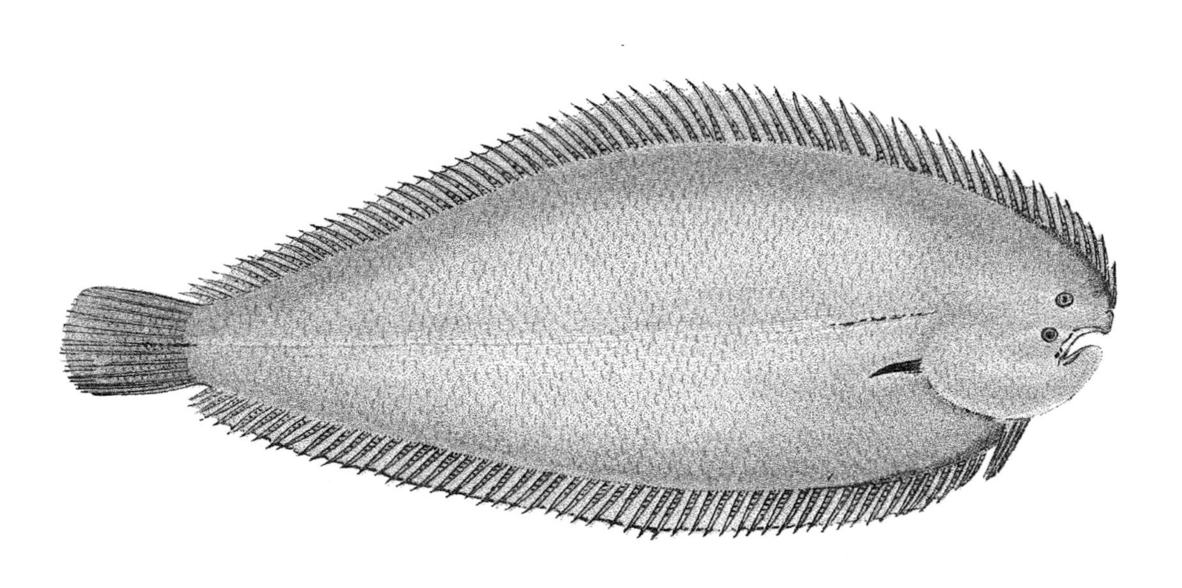

SOLENETTE.

CLXXIX

SOLENETTE.

LITTLE SOLE.

Solea parva, sive Lingula,	WILLOUGHBY; p. 102. Table F. 8, f. 1.
Monochirus minutus,	PARNELL; Magazine of Zoology and Botany,
" *linguatulus,*	vol. i, p. 527, pl. 16.
	YARRELL; Br. Fishes, vol. ii, p. 355.
Solea minuta,	GUNTHER; Cat. Br. Museum, vol. iv, p. 470.

THIS smallest of the Soles has always been known as a distinct species by fishermen, although disregarded by them as of little value for the market; and in consequence of this it remained only obscurely known to naturalists until it was noticed by Dr. Parnell, who found it in some abundance in trawls on the south coast of Devonshire. It is also common in Cornwall, where this method of fishing is practised; but I do not know of an instance where it has been taken with a line; which may be because it keeps at a good distance from land, where no hooks are employed of so small a size as could be supposed to enter its mouth. Sometimes also it is procured from the stomach of the larger fishes.

The length of the example selected for description was five inches, which appears to be the usual size; breadth one inch and five sixths, exclusive of the fins: the general proportions as in the Sole, but more tapering towards the tail. The forehead rises in a curve, but is less rounded than in the Lemon Sole. The mouth twisted, and large in proportion; jaws more equal than in the Common Sole; but the under jaw projects, and has prominent teeth, those in the upper side small. Eyes near each other, the right eye near the angle of the mouth, both sunk in the surface; nostrils close in front of the right eye. Head and body covered with ciliated scales. Lateral line straight, passing over the tail fin, which is round. Upper pectoral small, upper ray long and arched, the second about half the length of the first; no pectoral fin on the under side. The dorsal fin begins close above the upper jaw, and ends near the tail without being joined to it; and the same with the anal, which begins close to the ventrals, which are under the throat. General colour a faint pink; iris of the eye golden.

This species is described as sometimes not having the rudiment of a pectoral fin on the blind side.

END OF VOL. III.

B. FAWCETT, ENGRAVER AND PRINTER, DRIFFIELD.